the last mortal generation

*The writing is delightful, often thrilling, and never perfunctory
or condescending. The subject matter includes most of the topics central
to human concern, incisively interpreted. Whence, What, and Whither
ourselves and our universe? Much is asked of the reader, and much is given.
In the last few years there has been a flood of books interpreting science
and conjecturing about the future; this is arguably the best.*

Dr Robert Ettinger,
author of **The Prospect of Immortality**

D0841169

the last mortal generation

Generation

DAMIEN BRODERICK

HOW SCIENCE WILL ALTER OUR LIVES
IN THE 21ST CENTURY

NEW
HOLLAND

Published in Australia in 1999 by
New Holland Publishers (Australia) Pty Ltd
Sydney • Auckland • London • Cape Town
14 Aquatic Drive Frenchs Forest NSW 2086 Australia
218 Lake Road Northcote Auckland New Zealand
24 Nutford Place London W1H 6DQ United Kingdom
80 McKenzie Street Cape Town 8001 South Africa

First published in 1999 and reprinted in 1999

Copyright © Damien Broderick 1999

All rights reserved. No part of this publication may be reproduced,
stored in a retrieval system or transmitted, in any form or by any
means, electronic, mechanical, photocopying, recording or otherwise,
without the prior written permission of the publishers
and copyright holders.

National Library of Australia Cataloguing-in-Publication Data:

Broderick, Damien, 1944-.
The last mortal generation.
Bibliography.
Includes index.
ISBN 1 86436 440 8
1. Longevity. 2. Immortalism. I. Title
612.8

Project Editor: Howard Gelman
Designer: Peta Nugent
Typesetter: Midland Typesetters
Printer: Griffin Press, Adelaide

For Arthur C Clarke
Who profiled the future and
dreamed a future of advanced sciences
indistinguishable from magic

Contents

one: ageing – & how science will end it

No one but a crank would say that a cure for aging is just around the corner. Fortunately for the planet, progress on this last and most challenging of biological problems will be slow, giving society time to adjust to its consequences … But science is as full of surprises as nature, and progress often comes unexpectedly … Not all creatures are subject to senescence. Death is not a necessary fact of life, and rates of deterioration vary enormously … Late in the next century, our descendants may see the conquest of time …

Roger Gosden, Professor of Reproductive Biology, University of Leeds, in *Cheating Time* (1996)[1]

In 25 years we could see the creation of the first products that can postpone aging significantly. This would be only the beginning of a long process of technological development in which human life span would be aggressively extended. The only practical limit to human life span is the limit of human technology.

Michael Rose, Professor of Ecology and Evolutionary Biology at the School of Biological Sciences, University of California, Irvine[2]

Some time during the twenty-first century, ordinary humans like you and me—or our children or grandchildren—will be offered new medical treatments that will lead, eventually, to physical immortality.

Science and technology, to be more precise, will provide the option of indefinitely extended youthfulness. That will be the start of an epoch when old age is unknown and nobody ever again will *need* to die—except by accident, assault or deliberate choice.

If so, we ourselves, or our children, are the last mortal generation, the last doomed to age and die just from a quirk of evolutionary genetics. It's an irritating thought at best, a profoundly tragic one at worst. Think about it, roll it around for a moment on your mind's tongue: *we* might be the last humans in history doomed to perish simply because we can't yet do anything about it.

But suppose we are in luck. Suppose that longevity researchers like Michael Rose are correct, while observers like Roger Gosden are being unnecessarily cautious. (Professor Rose, who already has doubled the normal life span of one laboratory animal, declared that science 'will postpone human aging substantially in the future, doubling the human life span at least; when we have accomplished this we will be ashamed that we did not work on it much sooner.') If that is so, and if we as a community are prepared to fund the ongoing research that is yielding one breakthrough after another in the labs, here's the *really* good news.

You and I might be members, instead, of the first *immortal* generation.

I hope that stopped you in your tracks.

At Play in the Fields of Science

Four hundred years ago, Galileo Galilei timed a swinging chandelier against his own body's pulse and uncovered, in his blood's rhythms, the key—although it would not be turned until our own time—to estimating the age and extent of the very cosmos itself. In this book, I invite you to join me on a preliminary sunny picnic stroll, fortified by cakes and champagne, through the fields of gold spun by that miraculously effective group mind, science. No part of this expanding knowledge landscape will be untouched by the arrival of extended life.

As we proceed, we will scale—or at least scan from a distance—four crucial peaks. We'll gaze at those vivid Siamese twins so close to us, Life and Mind, and then peer into the strange mirrors of worlds small and large, the invisible picometre world of the Quantum and the great thirty billion light-year expanse of the Cosmos. Astonishingly, perhaps, the single great motor buried in the core of that landscape, running it all, the same motor that governs our mortality (and which we hope to redirect and cheat, to provide ourselves with indefinitely extended life) is blind evolution.

The collective mind we call science—fallible, contestatory, driven by ferocious passions and patient effort—promises to deliver us what religions and mythologies have only preached as a distant, impalpable reward (or punishment) to be located in another world entirely.

THE END-AND THE BEGINNING

In January 1998, researchers Woodring Wright and Jerry Shay of the University of Texas Southwestern Medical Center at Dallas, and their colleagues at Geron Corporation, announced a biological discovery that was instantly dubbed a 'cellular fountain of youth'. This announcement appeared in the respected weekly journal *Science*, published by the American Association for the Advancement of Science.

The researchers had found a way to extend the mortal life span of typical human cells cultured in the lab, using an enzyme, telomerase,[3] usually restricted to only a few of the body's tissues. (To be more exact, the cells were given an artificial gene construct coding for a part of the telomerase enzyme.)

A full six months after that epoch-making announcement, Wright told me that the treated cells continue to divide long after they normally would have perished from old age. The untreated control cells had reached senescence (in effect, standard fatal old age for such cells) at about 65 divisions. By July 1998, the treated cells, by contrast, had reached 180–200 divisions and were still alive. 'Thus,' Wright added, 'one could say they have now doubled-tripled their life span.' These, remember, are normal, healthy human cells. What's more, they show no signs of malignant transformation into tumorous cancers, usually the price paid by body cells that keep dividing beyond the usual shutdown point. In January 1999, cells were still healthy, at some 300 divisions.

Electrifyingly, Wright and Shay argue that normal ageing is fundamentally related to a simple flaw in the ordinary replication of the body's trillions of cells. A kind of cap, the telomere, protects the ends of chromosomes, the genetic instructions that guide the construction and repair of our bodies. Chromosomes are the twisty, artfully tangled strings of DNA and protein that comprise our design manuals in the core of every cell in our body. Their long, thin strings are built out of tens of thousands of genes linked together in a knotted line. Genes are both the recipe for every segment of the body, and part of the machinery for putting those segments together from the parts available in the world's organic stockpile. The flaw is that their telomeres[4] usually deteriorate with each cell division. We shall look at this in rather more detail in the next chapter. True, what Wright and Shay had found was not some *single* cause of ageing. As we'll see, ageing is a complex passage, with many interlocking causes, from healthy youth to decline and death. But the error they have repaired *is* a key flaw—and they have succeeded in reversing its ill effects, possibly immortalising cells without turning them cancerous. This is a spectacular power over death that has never been within human grasp until now.

The Bud in the Worm

Regular readers of *Science* would not have been absolutely surprised by Wright and Shay's news, because nearly two years earlier, in May 1996, they had already learned of an equally astonishing laboratory feat. The life span of a perfectly ordinary animal, a tiny worm, was not just doubled or trebled—but multiplied sevenfold!

This was achieved by mutating several genes that control the creatures' rate of metabolism. Nematodes are not complicated animals—built out of only 959 body cells compared to the hundred trillion (10^{14}) in the human body (although our cells make up only 254 different cell *types*)—but it's an extraordinary proof of what is possible in life extension.

It seems plausible that science will find a way, sooner or later, to tame such lab breakthroughs, turning them into a routine drug or treatment applicable to humans as well as worms and cells in glass. If so, we will have the means to reset or even disconnect the genetic clocks that cause bodies to age, and finally to die.

OLD BEFORE THEIR TIME

In April 1996, the month before the nematodes' fresh lease on life was announced in the scientific literature, the first gene specifically known to cause the symptoms of ageing in humans—*age-1*—was reported.

One of the most heart-rending medical disorders is Werner's syndrome, which causes its young victims to 'age' tragically fast. (Even more heartbreaking is Hutchinson-Gilford syndrome, or progeria, which kills its prematurely aged victims before they reach adulthood.) Hardly more than a thousand Werner's cases are known worldwide, four-fifths of them of Japanese origin and 70 per cent the result of marriages between first cousins. In adolescence, people with Werner's fail to grow alongside their peers. Hair greys and sheds, age spots mottle their skin. By their late twenties they resemble shrunken geriatrics. Arteries and heart muscles are damaged, and risk of cancer soars—although, curiously, their brains are unaffected. Usually they are dead, of what looks like extreme old age, before they turn 50.

Cells cultured from Werner's and Hutchinson-Gilford patients show an abbreviated life span, as you might expect. While these gene disorder victims do not literally die of old age, clearly their premature decay tells us something quite significant about the inheritance of senescence (the biologist's word for 'old age').

The 1996 paper in *Science* announced that Werner's is caused by a defect in just a single recessive gene on chromosome 8, of which the luckless victims possess two copies (rather as cystic fibrosis patients, victims of the most common recessive genetic disease, have one copy of the cystic fibrosis gene from each parent). It codes for what is called a helicase enzyme, which controls the unwinding of the DNA helix during replication. Loss of this key molecule prevents other repair enzymes reaching the coded sequences inside each cell and 'proofreading' and repairing them. An obvious implication is that this molecular repair system might be boosted artificially, although not simply by increasing the dosage of helicase (which, paradoxically, can be lethal).

The goal of such research is obvious: not just to repair the dysfunctional genetic complement of people with horrible maladies like Werner's and Hutchinson-Gilford syndrome, although that is a noble and immediate task, but to arrest in all of us the physical changes we know as 'ageing'. That would necessitate renewing the instructional coding that has been damaged by various assaults of time and evolutionary compromises in our chromosomes. Rejuvenated, we would be deathless and young—until we perish by accident, novel infection, or deliberate choice.

WHY DO WE AGE AND DIE?

Ageing seems so natural, so inevitable, that we find it very difficult to summon the nerve to question its necessity, to ask: *why does life work that way*? Is the slow ruin associated with age a planned feature, or a horrible bug in the system of life, a kind of design flaw—or is it just due to an accumulation of accidents and assaults? Can anything be done about it, any steps taken to halt or reverse it? Does it even make sense to imagine extending the length of our lives beyond, say, a century or so, 120 years at best? (A French woman, Jeanne Calment, who sold pencils to Vincent van Gogh in Arles, died in August 1997 at an attested 122.)

Right at the start of our enquiry into the prospects for scientifically induced immortality, we need to get one key point clear. *Ageing* is not the same as *expected longevity*. Obviously 'ageing' refers directly to the passage of years, but it's used as shorthand for the physical and mental decline found in every creature, to some extent or another, as the years pass. Longevity, by contrast, is the maximum length of time a member of a given species can expect to remain alive under the best possible circumstances. As companions to humans, household pets such as cats and dogs live very much longer than they ever could in the wild, where almost certainly they would be felled young by disease, accident or predation from animals that feast upon them. (Even an immortal can be eaten by a lion or get run down by a truck.) As pets, animals age more gracefully and enjoy a larger proportion of their possible span—but their maximum longevity remains what it always was.

Until very recently, we humans rarely experienced true ageing. 'Old John of Gaunt', in Shakespeare's *Richard II*, was only 58—not really old by our standards. (Shakespeare himself died at 52, and in supposedly glorious Elizabethan England general life expectancy was a miserable

35—but not, of course, maximum longevity, which was much as it is today.) As Leonard Hayflick noted in *Scientific American* in 1997:

> *Either you are already old, or the odds are better than even that you will become old. This statistic became true only 40 years ago. Aging is an artifact of a highly developed civilization. For more than 99.9 percent of the time that human beings have inhabited this planet, life expectancy at birth has been no more than 30 or 40 years. It is only after we learned how to avoid animal predators, massive homicides, starvation, most causes of accidents and infectious diseases that it has become possible for a substantial portion of the population of developed nations to grow old.*[5]

Even with the best care, both dogs and cats grow less limber, lose acuity in their senses, suffer tumours and other metabolic disorders, and die within about 20 years at most. There is reason to think that even if every illness were prevented and each cause of decline were alleviated (by hormone replacement, say), so that the physical decline of ageing were pushed ever further into the future, still, like our pets, we'd die at the maximum age known for hardy, long-lived members of our species: for humans, about 115 or 120.

That has been known since 1825, when the actuary Benjamin Gompertz found a statistical curve marking the intriguing fact that once we've passed 30 (safe from the ravages of childhood illness that killed so many infants in his day), the chances that we'll die double every seven years. That is, probability of death in humans increases exponentially (just as population growth does). Interestingly, if you manage to make it to 80, the curve flattens out again. The very old are especially hardy—which is how they happen to be the very old. Your chances of surviving don't plummet as savagely, even though your absolute maximum is still around 115 or 120.

In the wild, the curve looks quite different. Chance of survival follows a 'half-life' curve from the start (perhaps from conception, since there are huge numbers of natural miscarriages). For each consecutive interval of time, only half of a given generation remains alive. Their numbers are scythed by predators that eat them, murderous rivalries with their fellows, infections that amount to the same thing on a micro-scale from the inside, or accidents, insufficient food, extreme weather conditions. By using our brains and our superior communal resourcefulness, we humans have managed to drag up the 'half-life' curve common in nature, like an archer stringing a bow. The very best a society could manage would be a 'square' survival curve—and, decade by decade, we are approaching this maximum. In such a medically advanced and secure world, nearly 100 per cent of the population that managed to get born would remain alive and healthy throughout their entire expected life span—and then tumble off the twig as they approached the maximum, like a well-maintained machine that one day simply falls to pieces in a heap.

None of this explains *why* bodies have this tendency to fall apart, especially since many of our tissues seem quite good at repairing themselves. Let's look at some of the explanations that have been explored for the damage that attends the passing of the decades following birth.

THINGS FALL APART

So said the Irish poet William Butler Yeats, talking about the social order, and life, and love. 'Things fall apart, the centre cannot hold.'

It's palpably true, isn't it, one of the great verities of life? There just seems to be a tendency toward disorder and corruption built into things. Science has a name for it, when it is found within closed systems: entropy, the inevitable victory of noise over signal, melody and pattern. When you look into the physics and mathematics of any closely bounded section of the world, entropy's blight is due to a simple, terrible truth: there are very many ways for things to go wrong, but only a few ways for things to go right, and available energy and pattern are constantly being lost as self-defeating wrong turns are taken.

Without ceaseless effort at upkeep and repair, constantly topping-up inputs of energy and rectifying information, things do indeed fall apart. In the slick words of real estate agents trying to unload a lemon, the ageing human body is a 'renovator's dream'.

THEY'RE COMING IN THE WINDOW

My parents enjoyed a foolish chanted song that ran around and caught its own tail, until everyone fell about laughing at its silliness. 'They're coming in the window,' Mother chanted. 'Then shut the window,' Father suggested. 'Now they're coming in the door,' Mother pointed out. 'Then shut the door.' 'But they're coming in the window.' 'Then shut the ...'

It's just like that with the human body, as with any organism. The body is a sprawling gappy household in which family, neighbours, pets, farm animals and random beasts and birds of the fields ceaselessly tramp in and out, snatching scraps of food from the table, dumping supplies into the larder, leaving muddy tracks everywhere, bringing in their friends to stay for the night, occasionally victimised by infestations of pests and raids from crazed bikers living up the road. Sometimes there's a war and the place is completely overrun by people brandishing weapons and cryptic instructions. With a bit of luck you can get rid of them before the place goes up in smoke, and then you need to set about fixing up the blundering damage they've done—even if they were on your own side.

We age, according to this story, because each of us is a kind of colony of cells, an ecology, leaking information and materials in and out, many of them neutral or beneficial, some of them hazardous, some frankly lethal. Eventually the trespassers take over, and there goes the neighbourhood.

STICKS AND STONES ...

... can break your bones. So can falling down, like Jack and Jill, and breaking your crown. But so, too, can simply sitting still and waiting long enough. The calcium will leach out of your poor bones, and your joints will seize up, and you will get less and less limber, more and more likely to break—even at the level of the soft, squishy cells we're made of. From the micro levels of molecules all the way up to organs, limbs and the entire organised dance of the complete body-brain complex, we are prey to the assaults of sticks and stones. Eventually, the accumulation of this subtle or blatant battering wears us down and finally kills us.

POORFREADING ERRORS

'Poorfreading' is an old self-referential joke among those devoted souls who do eye-destroying proofreading for publishers, picking up the spelling mistakes, transposed letters and words, disorganised blocks of information.

Not only sticks and stones break your bones, after all. Names (especially the wrong ones) really *can* hurt you. That's true on a psychological level, and the degree of resilience with which you cope with nasty, stressful verbal attacks from others probably accounts for why some people wither up and die and others fight on grimly until they're ancient and toothless.

But 'names can hurt you' is true in a much more profound sense. The DNA instructions written in the chromosomes inside each cell— the genes that define our anatomy, most of our physiology (the way our body works) and some of our behaviour—are words in the DNA code. Like magical spells—if magic really did work in the way superstitious people thought it did—these words have to be uttered correctly, and in the right order, or two unpleasant things can happen. Either *nothing* will happen when you desperately need something quite precise to occur, or *the wrong thing* will happen and the body is in *real* trouble.

Every time a cell renews itself, there's a chance a copying error will occur in the three-billion-letter, 100,000-word book of human life. Luckily, powerful proofreading devices are also built into each cell, so most bloopers are swiftly picked up and set right. Not all, alas, because it's an imperfect world (the mark of entropy, once more). Some of the code errors are caused by mistakes of micro-chemistry. The wrong chemical module is plugged in by mistake, or the correct one left out. Others are due to random bursts of radiation (from space, from the radioactive potassium and other trace radioactive elements in our own bodies, from the earth beneath our feet, from the sun).

So bad words *can* hurt. Ultimately, the final sentence of death to which all of us are subject, however careful we are in avoiding accidents and living a wholesome life, might turn out to be a sentence botched by a DNA spelling mistake. On the other hand, why should almost identical proteins carried by mice and men be damaged by wear and tear at such different rates? As immunology professor William R. Clark asks: 'Why does one animal live three years, and the other eighty?' Perhaps, he suggests, 'death is genetically programmed, in the same way eye color or cholesterol level is'.[6]

RUNNING DOWN, WEARING OUT AND SHUTTING OFF

Experts in the medical and scientific fields that concentrate on ageing tend to get rather cross when senescence is called a disease. That's not just because we don't yet have a cure for such a disease—there are many disorders that still elude the power of medicine, but they are clearly specific illnesses. Ageing and death are not illnesses, in this view. They are just what happens to body and mind over the decades. They chart the normal course of life's arc.

Another way to look at this question of terminology is to ask whether falling out of a tree and breaking your leg is a disease. Of course it's not, although if you manage to get the wound infected by opportunistic micro-organisms you could end up with a diseased limb as well. What if you're run down by a bus while crossing the road? Not a disease, even if major organs are bruised or even damaged beyond repair. Most of the disagreeable effects of age are, as it happens, quite like a very slow-motion version of a multiple automobile pile-up on a busy freeway. Sometimes we are contributors to the carnage, sometimes we're just innocent bystanders caught in the explosive impacts. Either way, the slow, remorseless injury done to us over the course of a normal lifetime is closer to a series of billions of smashing microscopic impacts than to the internal ecological contests we mean by the word 'disease'.

Consider what happens to our bones and joints. One of the terrors of old age is the hazard of a simple fall. Children can tumble over and spring up without a bruise. The elderly are likely to burst blood vessels in their skin and crack their brittle bones like dry sticks. Joints—the knobby ends of bones, and the sockets they sit in—lose the protective fluids and cartilage that guard them as the levers of the body swing and rub. In youth, these shields are maintained, by and large, as abraded cells are replaced under the guidance of local genes. As we age, the cells grow forgetful and inept, and our joints and bones become perilously vulnerable to the natural shocks of daily use.

In some cases, this is only to be expected, given the way we're designed. Our genetic programming causes baby teeth to fall away as we reach toward and beyond adolescence, and mature dentition erupts from the gums. Civilised diets rich in sugars trigger a cascade of assaults on our teeth, and often we end up with decay that requires expensive dental work or extraction. Even if we keep our teeth in good condition, grinding food between them for six or seven or eight decades physically wears them down. And once they are gone, the genes are done with us—we get no backup sets of teeth in middle and old age. That isn't the case with some ruminant animals, however, whose teeth simply continue to erupt slowly as they are worn down, like a pencil lead you keep sharpened.

Similarly, it seems that brain cells do not replenish themselves. They can make new connections, which is how we learn, but once a neuron is gone, that's it. In fact, we are born with a massive oversupply, and a ruthless scything occurs in our early years as our environment pares away the tangle of possibilities in our brains and hones us into the adapted locals we grow up to be. That's one reason we can't easily learn to speak a new language 'like a native' after a certain age. We start as infants able to babble every sound a human mouth can utter, but we retain the neural control systems only for those phonemes we hear in use all around us, by parents, siblings, friends, other language users. Eventually we lose the ability to distinguish some sounds that other languages effortlessly hear as quite separate—the most famous example being the Japanese difficulty in telling apart 'l' and 'r', or some Londoners' problems with 'v' and 'th' ('Come in and meet my fahver.' 'I'd rahver not!').

But many other parts of the body do quite a good job of repairing the damage inflicted by the slings and arrows of daily living. Our skin is itself a huge, subtle organ weighing nearly three kilos, and it's constantly flaking off us and regrowing. The dry skin tissue we ceaselessly shed adds up over a lifetime to about a quarter our adult body mass. True, its quality decays as the decades set in. Below the surface, or epidermis, is the strong, elastic dermis, and below that is the hypodermis, a smooth layer of fatty cells. Each of these layers degrades over time as our repair processes start to shut down.

Replacement of the outer layer slows, outstripped by loss, so skin gets increasingly thin and fragile after 50. Collagen in dermis cells, the protein that gives tissue its warp and weft, locks into stiff cross-linkages. Fat cells at the deepest layer clump together even as they thin out, causing the blotchy look of old faces. Cooling fluids from sweat glands dry up, as do the moisturising oils from sebaceous glands. The pigment melanin also clumps into 'age spots', which in pale-skinned people leaves the surrounding areas paler than before. Loss of muscle tone allows skin and features to sag and wrinkle, dragged down inexorably by gravity, ever slower to recoil after necessary stretching.

Still, as long as we are healthy, uninfected wounds can often heal without any visible scarring and without compromising the function of the damaged anatomy. Not long ago I developed a cyst (a useless bubble of liquids and proteins inside an epithelial membrane) near the outermost joint of my left pinky. It was irritating at best, causing the tissues of my little finger to swell painfully, and potentially harmful. Eventually I had it excised by microsurgery—biopsy showed it was benign, I was relieved to learn—which involved a long, quite deep incision curving down into my finger, skin peeled back in a flap, the cyst's watery bag scooped out, the flap drawn back to cover the wound then very neatly stitched up and bound.

A mere week later I had a friend remove the bandage and replace the dressing— too tricky to do one-handed—and just six days later the stitches were snipped and pulled out. (Ouch!) The healing wound site looked vile, as if I'd been the victim of a very small but obviously demented axe murderer. And yet after another two weeks the Frankenstein scar had diminished to a pale red line curving halfway down my little finger. After a few days of localised peeling, the flesh was smooth and wholesome, and most of the sensation came back. Our bodies are quite remarkably good at healing injuries of this kind.

They are less adept with other kinds of damage. Nearly 20 years ago I fell and struck the base of my neck on the edge of a bathroom basin. Over the next few years, the impact site became increasingly sore and immobile, and I started getting headaches that seemed connected to this pain in my spine's cervical region, where the big lumpy joints of the spinal column support the skull above the shoulders. X-rays show that the protective inter-vertebral disk at this site has sagged, like a flattened cushion, allowing the vertebrae to press together. Nerves running up the spine to my brain are now easily squeezed by surrounding tissues, when they should be neatly braced and protected by the spine's complex structures of bone, muscle and cartilage. The dreary result is regular, recurrent pain and headache—sometimes fairly disabling—*and it's never going to get better by itself.* Our body's genetic programs don't make provision for repairing such commonplace injuries, and the impairment only gets worse as we age. Later, we'll look at the evolutionary basis of such heedless oversights.

EATING DISORDERS

Humans, like all other reasonably elaborate animals, are highly complex living structures that persist from birth to death by absorbing other highly complex living structures into their own bodies, destroying the integrity of those victimised structures in order to capture the handy preassembled microscopic sub-units and energy supplies they are made of. That's true even for vegetarians.

Ageing disrupts our ability to perform these predatory necessities. For hunter-gatherers, which is the lifestyle humans followed for most of our evolutionary history as *Homo sapiens sapiens*, the creeping decrepitude of age made it harder each year to pull your weight, whether you were a hunter tracking game and fish or a collector fossicking for tubers, fruit and tasty bugs. Worse, though, is the way the body itself gradually loses the power to process food efficiently.

For a start, as we've seen, the teeth can decay and fall out, be knocked out by accident or ritual, and even if they remain in place they can be worn down by sheer abrasion against the food eaten in a lifetime. Knives, boiling pots, ovens and other high-tech means of predigesting chunks of animal muscle and vegetable fibre spare us some of this damage, but emulsified gruel is just as likely to leave sugary residues stuck between the teeth, a rich ecological niche for decay-causing bacteria with their secretions of acid. But even when the elderly manage to chew their painful way through a decent meal, their digestive system is less able to process it. We can starve amid plenty.

The digestive tract is itself a complex system, built from dedicated modules linked in a series of feedback circuits. Chewed food, when swallowed, is already partly processed. In the stomach, it is churned in a lake of acidic gastric juices. All the varied fats, proteins, hormones and carbohydrates that made up your rare steak and vinaigrette salad are partially broken down and stirred into chyme. With age, the stomach's muscles weaken and grow less effective, and we lose perhaps a quarter of our gastric juice supply. Less than half the amount of protein-munching pepsin is released. Heavy meals are far less comfortable to digest for those older than 60.

Chyme floods down into the small intestine, a looped tube lined with spongy rippling villi that massage the gloop, extracting nutrients with the help of chemical secretions from the liver and pancreas and passing them into the bloodstream. (Luckily, neither the pancreas nor the liver are markedly impaired by ageing.) Each villus cell has a limited life span of its own; after less than a week it is renewed from a stem cell population at the bottom of the villus. Even so, after age 70 we find it hard to absorb calcium, which compromises the strength of our bones. Vitamins D and B_{12} are also less readily absorbed. If these functions could be improved, the elderly would make better use of their food.

Chyme passes on into the large intestine, or colon, where most of the water is drained out. Mixed with huge amounts of dead bacterial 'flora'—symbiotic organisms that regard our bodies as their ecology of choice, and pay their way by helping break down the food we've eaten—the residue moves to the rectum and is eventually evacuated. This part of the cycle worsens somewhat with age; hence the attention to the state and regularity of their bowel movement that the elderly often manifest. The kidneys, which excrete urine, start to shrink in early maturity, their internal structure of tubes and filters hardening and even dying bit by bit. Since the kidneys are the station for filtering poisons out of the now nutrient-laden blood, this is not a negligible loss.

People fortunate enough to live to 90 lose half their kidney function. On the other hand, people can live quite comfortably with a single kidney.

Notoriously, the bladder also starts to fail in late middle age. All through the body, connective tissues are stiffening and suffering from cross-linkage, and this impacts the bladder by making it harder to expand enough to hold more than a few hours' worth of urine, and then equally difficult to drain what's there. More commonly, it is caused by neurological impairment (especially in diabetics), or mechanical problems caused by prostate damage in men and prolapse in women. Ultimately, the elderly can find themselves plagued by embarrassing incontinence—quite simply, an inability to know when the bladder is about to let go and dump the lot. On the face of it, this is the least of the problems facing people with a high and imminent risk of heart attack, cancer, influenza and a dozen other potentially fatal conditions. But for a social, urbanised animal like us, the consequences in humiliation and simple infuriating loss of command over oneself can be deeply humiliating and destructive. Freud was not entirely wrong about the gravity of toilet training.

THE ILLS FLESH IS HEIR TO

Physical change is not automatically bad for us, of course. The transition from infancy to childhood and adolescence, and then to full reproductive maturity, is mapped in advance in our genetic programming. The exact way it finds expression is somewhat flexible, responsive to the cues we encounter in the world—some from nutrition, some from the people we interact with. The arc of our life rises, plateaus and falls, from the feeble if lusty condition of the newborn through a significant watershed called the climacteric in middle age (historically, a phase few humans ever reached) and thence into the jolting decline toward inevitable death.

However joyous our life, there is pain and discomfort at each point in this arc. Babies scream with the frustration of not being able to make their wishes felt more precisely, because they do not yet have access to language. Teeth cut their way up through the tender flesh of gums.

Children bang into things as their developing brains learn how to coordinate an immensely complicated body in a world of rushing and often mystifying things and people. Adolescents suffer physical aches as bones stretch and muscles put on mass, as hormones change their programmed balance in the body's economy and the skin goes crazy with boils and scabs, as sexual characteristics erupt from the smooth skin of childhood and mark both body and mind with new agitations, confusions, hungers.

Women go through more of these staged alterations than men do, or at least those changes are more obvious and definite. The arrival of menstruation is taken in every human culture as the mark of womanhood, even in the many cases where it is hidden away as something shameful. Its departure thirty or thirty-five years later is equally momentous: the close of the option to bear children, the opening into a final reproductive phase as an elder, perhaps as a grandparent to her children's children, perhaps as a custodian and guardian of her culture's experience and wisdom—or perhaps, all too often, as a discarded leftover, ignored and allowed to decay and die by a careless society. Plainly, each of these stages is driven by biology and modulated by culture. At their base is a sequence of biochemical structures and operations that we are finally beginning to understand, and even to manipulate to our benefit.

The most drastic manipulation of human reproduction, short of germ-line genetic engineering, is cloning a person. The first publicly acknowledged preliminary experiments on human cloning were claimed in December, 1998, in Seoul, South Korea (but other specialists were quick to throw doubts on the claim). Infertility clinic researchers at Kyunghee University Hospital in Seoul said they used a method pioneered in Honolulu, involving the injection of a mature body cell nucleus from the mother, in her 30s, into an emptied, unfertilised egg. The embryo divided into four cells and was then terminated.

Our bodies and their workings are organised by a kind of biochemical cascade. Each of the organs and crucial life-sustaining systems—liver, kidneys, heart, skeleton, muscles, gut, blood, lymph, immune system—has its specific boundaries and monitors, even while they seamlessly interact with each another. The cells of the heart, or even different sectors of the heart, are quite distinct in shape and function from those of the brain or the soles of the feet. Yet the nucleus deep inside each of those cells contains exactly the same DNA recipe. Most of the body-making message is normally switched off or masked, which is why each cell type looks and acts rather unlike its neighbours in an adjacent organ. Nevertheless, they are coordinated by ceaseless flows of nutrients and transmitter substances. Some of these are hormone proteins that act directly to start, modify or stop a specific process. Others are neuro-transmitters that prompt and convey the electrical information that zips through the kilometres of nerves linking brain and body.

Brain *and* body? That is a somewhat misleading distinction, since of course the brain is part of the body. Still, it is the portion we particularly identify with the self. If limbs or even most of the torso are badly damaged, you remain you, but if the brain in your skull perishes you are gone, even if the rest of the body survives in a cradle of advanced machinery. And there is a designed division between the two called the blood-brain barrier, a defensive system that prevents large molecules from passing in the bloodstream from the rest of the body into the exquisitely sensitive lobes of the brain. Even so, we are becoming increasingly aware of the perpetual swirl of information from the brain to the rest of the body and back. So it should not be surprising that mood—governed to some extent by cognitive states—can influence immune responses, just as illness and pain can cause our thoughts to drift down dark corridors in sympathy.

MENOPAUSE

We imagine sex as the manifestation of the body *par excellence*. Not quite. The brain drives our sexuality. The reason women's bodies alter so drastically during menarche and menopause comes down, finally, to cir-culating levels of chemicals exuded by the pituitary gland deep in the brain, chemicals which in turn activate the production of other powerful chemicals or turn their secretion off. Luteinising hormone (LH) and fol-licle-stimulating hormone (FSH) flow in the blood from the pituitary to the ovaries and stimulate the reproductive hormones progesterone and oestrogens such as oestradiol, which help maintain tissue integrity. Menopause marks a change in a body's response to these pituitary messages, largely because a woman's ovaries are finally depleted of eggs. Unlike men, who keep producing spermatozoa all their lives, girls are born with a limited egg supply. Once all the eggs are gone, the ovaries can no longer respond to LH and FSH, even in elevated quantities—so other hormones necessary for health, chemical messengers formerly made in those tissues, simply dry up.

The fact that ova are a finite resource is a reflection of the evolutionary history of human ageing. A human female foetus contains a startling seven million eggs, which are screened and shed in vast profusion during the course of her mother's pregnancy. At birth, the number of ova in the newborn girl child is down to a million, and the number keeps dropping through life—a quarter of a million at puberty, only 25,000 by around age 40. When a mere thousand or so remain, the subtle feedback between pituitary and ovaries starts to shut down.

It has been suggested that surgically removing and freezing some of this wasted ovary tissue, transplanting it back in late middle age, might serve to extend the hormonal protection granted to the reproductively able. Indeed, experiments have shown that implanting ovary tissues from aborted female foetuses (which contain, as mentioned, a huge supply of egg cells) can extend an adult's sexual window. This is regarded by many as a repugnant option, and has been banned in some countries, but the point remains—it is the loss of eggs that activates the doleful cascade of menopause, not the other way around.

Once that happens, then, a menopausal woman's reproductive hormone system starts to shut down, and the long-term consequences for the whole female body are wellnigh catastrophic. As the least of her problems, fat is laid down more readily in some places and less so in others, so an older woman's body is likely to thicken and sag even as her features hollow and wrinkle. Her bones grow brittle. And, for a time, 'hot flushes' suffuse the whole body as blood suddenly rushes into a woman's surface tissues, trying to cool a body confused by the abrupt drop in levels of oestrogen. Luckily, well-tailored hormone replacement therapy can compensate for many of these uncomfortable and injurious side effects.

So is menopause a selected, organised change in the human body's dynamic, like the growth spurt and onset of sexual maturity in adolescence?

Or is it something more alarming—a simple failure of one life-support system after another, driven by the loss of ova, the exhaustion of the ovaries, and the consequent failure of all the reproductive machinery? Is the female human body coded genetically to 'degrade gracefully', as an engineer might put it, or is post-menopause life just whatever happens after the final page of the manual is turned and nothing more is written?

Hormone replacement therapy already shows that we can stall some of these uncomfortable and damaging changes, although sometimes at a price. As we'll see in a later chapter, evolutionary biology provides various answers to this puzzle, not all of them mutually consistent. One way or another, this event in the human life cycle is key to any successful understanding of age, senescence and the chances of contriving a medical cure for both—a cure, so to speak, for death.

In men, too, middle age (these days, late middle age) is marked by hormonal tides and ebbs. Nothing quite as drastic as menopause occurs, but circulating levels of sex hormones clearly drop—allowing many men, for the first time in their lives, to get a decent grip on their emotions. Reproductively, the impact is not as devastating. Within a few years women lose the ability to shed fresh ova, and those they do release during the dying phase of their cycles risk being impaired. Men, by contrast, can continue producing spermatozoa and the semen that sperm wriggle in, but the viability and number of those sperm also tend to decrease. As sperm density drops, the pituitary instructs the testes to increase production, but those orders have no effect, rather in the way a woman's ovaries ignore orders transmitted in the form of FSH. What's more, the prostate often enlarges, with an increased risk of cancer. The connective tissues and blood vessels in the penis become less flexible, and erections are more difficult to attain and sustain. The explosive popularity of the drug Viagra shows how significant this erosion of male potency has been to those afflicted, and provides a clue to the success of more general remedies for ageing.

Losing Your Mind

Because we regard the brain as the key to what once was called the soul or the spirit—that is, memory and continuity of your identity—

crumbling of the brain itself is the most distressing feature of growing old. As we've seen, attrition of the brain's billions of neurons (and other support cells) sets in very early. During infancy and childhood, hundreds of millions of brain cells perish as we are shaped by experience. That much is beneficial. The downside is that brain cells generally do not repair themselves, and they continue to die off. The horrible affliction of Alzheimer's disease looks like an extreme form of this decay, as plaques of amyloid protein clog up the brain's tissues, and the delicate lacework of the neurons clumps into tangles, leaving quite literal gaps inside the head that can be imaged using scanners.

Different regions of the brain, oddly enough, discard neurons at different rates. The brainstem, which controls heartbeat, breathing and other 'autonomic' regulatory activities, along with sleep and alertness, is a kind of hardened system proof against ageing—except for the locus coeruleus, linked directly to higher centres: two-fifths of its cells are lost by the time we are 65. A quarter of our Purkinje fibres in the cerebellum, crucial to motor control, vanish abruptly around the same time, which helps explain why older people suffer impaired coordination. The visual cortex loses as many as half its youthful neural quota by age 70 or so, as can the auditory region in the temporal cortex. Intriguingly, neither the high-level prefrontal cortex nor the emotion-controlling thalamus loses any significant neural structure (unless disease intervenes).

Luckily, then, even the oldest healthy human still retains a neural network sufficiently complex to sustain reasonable alertness and continuity with his or her earlier self. But if life span were to be extended greatly, not to mention indefinitely, this decay curve does raise the gruesome prospect of youthful multi-centenarians almost literally without a brain in their heads. Of course, by the time medical solutions are found that allow drastic life extension, these other hazards might well be soluble as well—or, indeed, mended by the very methods used to defeat systemic death.

At present, it seems that many potential problems arising from neuron die-off are averted by an equal and opposite neural plasticity. We can, to some degree, reprogram the physical structure of our brains. Obviously this must be true to some extent, because we keep learning new things and reorganising old knowledge and feelings, and this plainly requires our brains to grow new connections (synapses) between storage areas and allow others to fade away. In a computer, this could be done simply by changing the contents of a RAM chip or a hard drive. In a brain, built on quite different biological principles, it requires chemistry at the neuron level and some actual physical anatomical changes in the connectivity of the neurons that hold our memories.

THE HEART OF THE MATTER

Although the western world has been in the grip of a kind of hysteria about the threats of cancer and AIDS (truly horrible disorders that might be close to being solved), these two are by no means the major cause of death in human societies. Far from it. In poorer communities, of course, not all of them in the Third World, ancient curses still kill vast numbers of people, many in infancy—starvation, polluted water or none at all, curable diseases that have not been treated. In our own technologically advanced societies, however, heart disease takes out a whopping 43 per cent of those dying at age 65 or older. Cancer accounts for only 16 per cent. Stroke (cerebrovascular disease) kills about 10 per cent. Pneumonia and influenza, by contrast, account for about 5.6 per cent of the elderly. AIDS, it's tragically true, like motor vehicle accidents and suicide, typically kills its victims in youth, but the incidence is still very low, and falling in the western world. What is it about the heart which makes that crucial organ so vulnerable?

The heart is a muscle that works ceaselessly, taking in depleted blood from veins threaded all though the hungry tissues of the body, passing it to the leafy oxygen-rich pastures of the lungs, and returning its rich red harvest through the arteries and back to the body's suffocating tissues and organs. Problems with a system like that are inevitable. The muscle-lined supply tubes in and out of the cardiac system grow stiff and hard over the years just as other tissues do. Injudicious diet can encrust them with garbage, notably cholesterol and low-density lipoproteins, reducing blood flow. Calcium builds up in the arterial walls, and collagen cross-linking stiffens all the endlessly flexing parts. Poisons of one sort or another can damage the cells, hampering their capacity to work and repair themselves.

By age 65, we lose a third of our aerobic capacity—the maximum amount of oxygen we can take in during exercise. Tissues throughout the body slowly and chronically get starved of oxygen, and repair systems in the body start to cannibalise under-used muscle mass. The heart's left ventricle, which pushes oxygenated blood into its aorta and on out into the body, can thicken over the years if disease is at work, obstructing contraction, so the amount of blood expelled diminishes. As always, damage cascades. Heart attack (myocardial infarction) is literally the death of a portion of the heart, which can lead, of course, to the rapid death of the rest of the body. Heart failure, a different ailment, is reduced mechanical efficiency which blocks blood flow, and causes abnormal distension of the veins leading into the heart.

In short, numerous problems can assail the cardiac muscle and its pipes—deposits of fat in the wrong place, auto-immune injury arising from the body's over-eager response to infection by streptococcus (rheumatic fever), congestion, loss or interruption in the heart's regular beating (remediable these days by tucking a pacemaker inside the chest). Some of these potentially fatal nuisances can be offset by deliberate lifestyle change: giving up smoking, eating fewer saturated fats, walking a few brisk kilometres five days a week, reducing salt intake. These sensible options return us to something like the routines imposed by nature on our hunter-gatherer evolutionary ancestors, except that we retain the benefits of refrigeration, libraries and the Internet, rapid transport, contraception, antibiotics and a thousand other boons not known on the African and Asian plains in antiquity.

Even so, we are at the mercy of metabolic and genetic errors, thudding damage, and programmed shutdown. If we wish to get serious about defeating death and finding a way to remain young for as long as we wish, we're going to have to make some major modifications at the cellular and probably the genome levels. That is where wonderful discoveries like the Wright-Shay work on telomerase promise to revise everything we have ever known about life and the supposed inevitability of death.

two: death – is a cure possible?

In the morning they are like grass which groweth up. In the morning it flourisheth, and groweth up; in the evening it is cut down, and withereth.

Psalm 90:5

All flesh is grass, and all the goodliness thereof is as the flower of the field. The grass withereth, the flower fadeth: because the spirit of the Lord bloweth upon it: surely the people is grass.

Isaiah 40:6–7

Nothing can be said to be certain, except death and taxes.

Benjamin Franklin (1789)

Our tour of the ways we fall apart over the decades, ghastly if extremely abbreviated, has left one all-important question hanging in the air: *why does it happen in the first place?* We often don't see that this actually *is* an important or even meaningful question. Yet death snatches at some creatures within a day, while others endure for millennia. All flesh might well be grass, but some vegetation certainly isn't. Sequoia trees can live for 3000 or 4000 years. Bristlecone pines in the Californian Sierra Nevada are more than 5000 years old and remain fertile. Certain creosote bushes are thought to live for at least 10,000 years.

Indeed, it is possible that some fish, birds and land animals can escape death until accident, disease or predation slays them. Writing in *The Journal of Gerontology: Biological Sciences* in July 1998, one of the world's experts on ageing, Caleb Finch of the University of Southern California, mentions 140-year-old orange roughy, Northwest Pacific rockfish, in which ancient females possessed 'abundant, newly formed eggs and ... appeared in fine health'. Other creatures continue to grow larger but show no signs of ageing, and others again just mysteriously plateau. Robert Gosden's *Cheating Time* contains an intensely poignant series of three photographs, showing Scottish ornithologist George Dunnet with a small white and grey fulmar petrel he ringed and tracked across 45 years. In the first shot, taken in 1950, Dunnet is a handsome young fellow of about 30. In 1976, the scientist's hair is receding, and his face is starting to sag under the forces of late middle age. The bird is unchanged. By 1995, the year in which Dunnet died, he is plainly an old man—and the bird still looks exactly the same. Despite such ordinary evidence for human mortality, Finch concluded in 1998 (drawing on those extraordinary indications of some non-human freedom from its grasp) that no firm lifespan limit is necessarily built into the human genome.

Ben Franklin's resigned aphoristic cynicism is both understandable and somewhat misleading then, and as we shall see, so too are the biblical laments. Actually, of course, levying and paying taxes are both deliberate social choices. Most of the cultures throughout prehistory would have been astounded at the idea of mandatory, graduated tax. Perhaps it will prove to be so with death also. Perhaps we do not fall because 'the spirit of the Lord bloweth upon' us, but because of evolutionary short cuts that intelligence can improve upon. 'Aging came in by the back door,' as Roger Gosden states bluntly, 'uninvited but unopposed.'[1] So is physical, earthly immortality a realistic hope? 'Yes,' as the transhumanist Jeff Dee put it aptly, 'so long as the definition of immortality is "lacking a pre-programmed life-span" and not "total immunity to destruction".'

Let's pursue Franklin's once-amusing aphorism for a moment. A cynical view of taxation in the modern world might regard it as an unholy blend of nobility and theft. 'Tribute' is a very ancient practice of thuggery, enforced by the strong and clever, today given a more acceptable spin, usually by wrapping itself in a flag. Yet taxation is also a way for all, to the extent that they can afford it (or the extent that their tax lawyers can't find evasive loopholes), to contribute to great large-scale public goods. Many massive projects aren't easily managed on an individual basis: road construction and traffic signalling, some measure of health services for the poor, the largest-scale provision of law, government and military (although here, too, not all the alternative methods have yet been tried).

Whichever side of the balance you care to stress, it's clear that humans could do without taxation if they were prepared to give up the security (and dread) of living in a modern hypercivilisation. Death, on the other hand, seems perpetually with us, the necessary concomitant of life.

In fact, it's not. It is no more than a strategy adopted (unthinkingly, of course) by certain lineages of living creatures hundreds of millions or even billions of years ago. Death had advantages and drawbacks, just as taxes do. And once a species has moved in that direction, it is as difficult to reverse the decision as it would be to secede from the taxation requirements of your local polity.

THE STRATEGY OF INDIVIDUAL MORTALITY

Many creatures are deathless, to all intents and purposes, as they swarm happily over the earth, and in the sea, and deep beneath the ground. Individual cells all die, of course. Every replicating cell has its own life-span limit, unless it's a cancer cell, and then it will perish, at the very latest, when its host victim is converted to nothing but tumour. But an individual can be defined by its characteristic and unique DNA code. Creatures that reproduce by binary fission—simply doubling their chromosomes in an engorged bath of cytoplasm, before splitting into two genetically identical 'daughter' cells—can be regarded as immortal. Sooner or later, any individual member of the cellular clone will stop replicating, and that will be the end for it. But its separated twins and offspring will keep the line going until random mutations, or genetic material scoffed down and assimilated from other creatures, rewrite their identity code.

But is that so very different from us? We retain our core genetic identity from conception to death, although clearly we do not stay utterly the same over a lifetime. Various cells in the body do mutate as a result of radiation, chemical attack and other hazards. More to the point, we can learn from experience, redesign our responses to the world, morph our personalities and memories. That's one of the benefits of being a complex, multicellular organism. Still, we do regard ourselves as basically stable at some core level. We are *selves that persist*. Maybe the lineages of bacteria and other asexual creatures share something of the same mutable identity down the centuries.

What we and bacteria have in common is a constant risk of cellular and even genetic or program-control damage. That's the main cause of ageing, after all. Things not only fall apart, they lose the capacity to put themselves back together. The mystery is why that second part should be the case. We humans are an exquisite, harmonious composite of 60–100 trillion different cells of more than 200 different kinds. Each of them contains the same fundamental genetic information, although it is differently masked in each kind of cell. An engineer might suspect that when one cell gets damaged there would be some way to use message redundancy to correct the error.

CHECK YOUR MESSAGES

Redundancy is how a very weak signal is received from a space probe in the depths of the solar system despite a blizzard of interstellar noise. It works this way: the same message is sent many times, and only the common, overlapping parts of the received signal, shared by a majority of the repetitions, get accepted as part of the final decoding. While each transmission might be flawed by random blurts of cosmic noise, it's very unlikely that the bits of the message stream that happen to be damaged will overlap from one transmission to the next. Couldn't the body use this principle to check on the reliability of its many copies of the genome? At worst, couldn't an organ simply *discard* any errant cell, and copy a replacement from a neighbour with an uncorrupted identity template?

One reason evolution never tumbled to this kind of repair technique is that evolutionary change itself depends on code errors. Most are neutral or lethal, it's true, while only a few prove beneficial to the animal bearing such mutated or altered genes. Still, given a sufficiently large number of modified animals, and a long enough period of time for them to compete, survive and prosper or die without progeny, certain helpful variants will slowly dominate their species gene pool. Novel variants can even skew the local population so drastically that an altogether new species is thereby created. A species with perfect remedies for code errors would become stagnant, slowly surpassed by its more fallible competitors. Intelligence, of course, would permit humans to grow and adapt *without* the terrible cost of enforced mortality.

Unfortunately for us, then, evolution never hit upon the redundancy solution to the problem of memory loss at the cellular level, although the immune system uses principles not unlike that to identify, mark and delete damaged body cells and invaders alike. Some day, probably in the twenty-first century, we will learn to do so technologically, using a swarm of fantastically small programmed nano-machines, built largely out of diamond-configured carbon, to eradicate badly malprogrammed cells and keep the rest up to the mark.

'The typical medical nanodevice,' according to Robert A. Freitas, 'will probably be a micron-scale robot assembled from nanoscale parts. These parts could range in size from 1–100 nm (1 nm = 10^{-9} metre), and might be fitted together to make a working machine measuring perhaps 0.5–3 microns (1 micron = 10^{-6} metre) in diameter. Three microns is about the maximum size for bloodborne medical nanorobots, due to the capillary passage requirement.' Nanomedical treatment to fight infection will require 'an injection of perhaps a few cubic centimetres of micron-sized nanorobots suspended in fluid .The typical therapeutic dose may include up to 1–10 trillion individual nanorobots, although in some cases treatment may only require a few million or a few billion individual devices to be injected.' Such machines would not be self-reproducing, as are bacteria and even viruses (which use borrowed cellular machinery). They would be monitored and constrained from outside the body by specialists using scanners, so they would not easily run out of control.[2]

Unlike medicine—today's variety, or tomorrow's nanomedicine and nanosurgery—there's no goal, no intention in Nature's long-term methods, just a simple statistical truth that spendthrift variation and remorseless death of botched animals will yield novelty in the long run. It is likely that creatures bucking that current by maintaining the same genome intact over hundreds of millennia, let alone millions or billions of years, would prove unequal to the contest for survival. Such relics would be legacy organisms, superannuated at conception. And they would simply not persist. If we humans do decide to adopt this path, or a version of it, we will be protected from obsolescence by the keen brilliance of our accumulating knowledge, our vast cultural memory, our imagination and ability to forecast possible exigencies more accurately than blind natural selection was ever able to manage. After all, more than 99.9 per cent of all the species that ever lived on this planet are now vanished.

Forgetting to be Young

In a sense, then, ageing and death are akin to a failure of memory. Generally, since our memories ebb as the years pass, we think it's the other way about—that memory fails with age. Oddly, that's not quite true. Most people find that they lose short-term memory first. They cannot recall what they were just about to do, or who this person is who has arrived to deliver their Meals on Wheels, yet they nod over richly hued recollections of childhood and buoyant or difficult young adulthood. The difference between long-term and short-term memory is a function of the brain's modular structure, which we'll return to in a later chapter. For now, we need to see that ageing is itself a kind of memory lapse.

Things fall apart on many levels. If your arm is blown off in war, or your immune system hopelessly damaged by a retrovirus, part of your developed self is gone for good. That's due to the way in which complex animals manage their economies. We start with two cells containing not much more than half a complete human genome (plus some factory machinery and a prepared lunch), which blend into a single new unique human cell. Unlike any other cell, except for the botched and monstrous things that turn into proliferating cancers, this fertilised egg has the capacity to double, redouble, double again and yet again, into a clump of genetically identical cells. But those cells are *not* strictly identical, because they are positioned differently in relation to each another.

The moment one of them is selected as the character at the front of the crowd, the cell mass has the opportunity to differentiate, to start specialising. Spatial and chemical gradients begin to draw the wildly replicating cells into a preprogrammed shape, and as the shape emerges the cells in each location begin to concentrate on just one kind of duty, to lose their total potential. Portions of their DNA ensemble are being masked, other portions forced forward so that their coded formats can be expressed as region- and role-specific proteins. This cascade of differentiation is what builds us, but also what prevents us from regrowing lost limbs and organs.

Despite a remarkable degree of local recovery from limited injury—not to mention the routine wear and tear of living—cells individually and in aggregate simply can't reverse their specialisation and start afresh with the totipotent gusto of an early embryo cell. Actually, there is one known exception to that blanket prohibition—physically removing the nucleus of an adult, differentiated cell and fusing it into an embryo cell lacking its own nucleus. This method brought us Dolly the cloned sheep, plus several Japanese cloned lambs two years later (although they quickly died). Another method, reported in *Nature* in July 1998, yielded wholesale cloned and recloned lines of mice in Honolulu, using a micro-injection technique with a success rate of three per cent—still low, but better than earlier methods. So far, though, we do not know of any means to prompt the regrowth of a hand from the stump, or a fresh liver in situ, in the way a few animals like starfish can rebuild a missing arm.

So damage is not lightly reversed, not just at the large scale of fingers, eyes, hearts and bald heads, but right down at the cellular level and beyond. The breakthroughs of recent years that make the prospect of immortality (or at least massive extension of youthful longevity) more than just a wistful dream derive from our new understanding of just what kind of damage this is, and some of the means by which its unpleasant effects can be averted, minimised, or even revoked.

THE INCREDIBLE SHRINKING GENE

The genes' DNA is based on just four uncomplicated chemical units, coded as T (for thymine), A (adenine), C (cytosine) and G (guanine).

We now know the chromosomes are capped at either end by disposable non-gene chemical gadgets, like snap-off sections of a chocolate bar, which are themselves covered by a kind of cloak that also shrinks with repeated replications. These *telomeres* are just repeated stretches of a six-letter word built from three of those coding units or bases, T, A and G. Spelling out the non-genetic 'word' TTAGGG (or T_2AG_3, to those in the trade) and repeated thousands of times at a stretch, these rather inert telomeric chunks of DNA 'alphabet' help protect DNA stability—for as long as they are intact, or for as long as there are sufficient to do the job. Generally, though, fewer are copied with each replication in the cell's repair cycle.

We start at conception with some 15,000 non-gene bases in this 'terminal restriction fragment' at the ends of our chromosomes, which is whittled by birth to about 10,000—half of them in the telomeres themselves, half in the adjacent subtelomeric region. That leaves us with about 1700 TTAGGG units per chromosome tip to get through life with. (The average gene is 25 times as long as your telomeres at birth, and the entire chromosome is 25,000 times as long.) By age 70 or so, you'll be down to an average of less than two kilobases per telomere in replicating cells, since each time a cell in your body divides to replace itself, the telomere tips tend to shorten. They do not exactly run out, but more and more cells end up with too few to launch a reliable replication cycle. That *doesn't* happen to germ-line cells (those producing ova and spermatozoa) and a few other kinds of cells that need to proliferate to keep us alive: stem cells in the bone marrow, active regions in hair follicles, intestinal crypt cells. Special machinery is in place to guard the crucial sex and specialised cells from telomeric deterioration. And the tragic disorders mentioned before are now known to exhibit pathologically shortened telomeres (although they might not cause the disease).

HITTING THE HAYFLICK LIMIT

Take a human cell, a fibroblast, derived from a newborn baby or its placenta, place it in a nutrient culture on a glass or plastic Petri dish, and it will divide up to 90 times. Then it lapses into senescence and eventually perishes, having reached its 'Hayflick limit'. When this replicative boundary was found and described nearly 40 years ago by Leonard Hayflick, Paul Moorhead and their associates at the Wistar Institute, it came as a great shock to cell physiologists. Earlier researchers, notably Alexis Carrel, a Nobel medicine and physiology laureate, had taught that ordinary cells grown in a culture would remain alive indefinitely. Not so. Worse still, a cell from a person of 70 has even less kick left in it—it will stop replicating after 20 or 30 divisions. (And yet the healthy sperm produced by an elderly man, while less abundant and more likely to be damaged, remain just as 'youthful' as those of a man in his prime.)

Could the Hayflick limit be largely a by-product of telomere shortening? Might cells estimate their permitted longevity by checking how much of the cap remains? The TTAGGG regions form a kind of neutral runway or docking station allowing the protein copying mechanisms to attach themselves to the chromosome. Once telomeres are sufficiently reduced in number of TTAGGG units (Hayflick refers to them as an 'event counter' or 'replicometer'), there's nowhere for the molecular machinery to dock, wrecking the cell's 'proliferative capacity'. But telomeres can be rebuilt. Wright and Shay managed to get their healthy human fibroblast cells to continue dividing indefinitely by introducing the repair enzyme telomerase.

TELOMERASE TO THE RESCUE

Once the original telomeres are significantly depleted, cells can no longer divide unless the caps are rebuilt. The tissues built from those telomere-depleted cells have become senescent. All the genetic information is still there inside the helical coils of DNA, but the machinery of the cell can't access it to allow replication, although even in senescent cells some genes at least can be read and transcribed. Telomeres, in this sense, can be thought of as an essential password to the software of the cell's nucleus.

That much has been suspected since 1988, when Robin Allshire at the celebrated Cold Spring Harbor lab read the human telomere sequence. Australian-born Elizabeth Blackburn, now professor and Chair of the Department of Microbiology and Immunology and professor in the Department of Biochemistry and Biophysics at the University of California, San Francisco, had already decoded telomeres from tiny *Tetrahymena*, a pond organism. With her student Carol Greider at the University of California, Berkeley, Blackburn had first identified telomerase in 1980. (For that feat she won the prestigious 1998 Australia Prize for outstanding achievement in molecular genetics promoting human welfare.)

In 1990, Cal Harley, Bruce Futcher and Carol Greider showed in a *Nature* paper—'Telomeres Shorten During Aging of Human Fibroblasts'—that telomeres really are the cell's meter. The more recent breakthrough is this: using that special enzyme, telomerase, it is now possible to repair the telomeres of chromosomes in cells that don't possess it naturally. (They all still have the genes to make telomerase, but in most cells it is switched off—probably as a precaution against cancer, as we'll see. In fact, most cells keep most genes turned off most of the time, switching on only those relevant to their own special kinds of tissue, whether these are skin, heart, brain, gut or whatever.)

Telomerase is what's called a ribonucleoprotein complex, comprising an RNA template for the synthesis of the repeated TTAGGG sequence, plus a reverse transcriptase—an enzyme allowing that RNA message to be read like a mirror image back into the chromosome's normally inviolable DNA. Blackburn has called the enzyme 'a crude little copying machine. It's thought to be one of life's ancient relics which got fossilised into our cells. Had it been useless, it would have been selected out ... But while it may appear cobbled together, it's a system that works.'[3] By adding telomerase, Wright and Shay managed to force healthy human cells to keep dividing in their Petri dishes, which have now doubled or trebled the cells' lives. There seemed no limit in sight.

Oddly enough, Hayflick, now professor of anatomy at the University of California, San Francisco, had declared as recently as the 1996 edition of his book *How and Why We Age* that 'it might be ten thousand years or more before the maximum human life span reaches even 120 years.'[4] It is notably encouraging, then, that within two years he was hailing Wright and Shay's technical breakthrough: 'This is a monumental advance in the understanding of the molecular genetics of aging. The telomerase gene will likely have many important applications in the future of medicine and cell engineering.'

We need to maintain a grip on our understandable hopes. In mid-1998, Blackburn told me, a paper was published by researchers at the Cold Harbor Spring Laboratory 'in which they have repeated the experiment using cells from a human tissue cell type different from the type used in the originally published experiment of extended life span. They now find that these other cells do not have their life span extended when the same experiment of expressing the telomerase gene is done.' That paper, 'Myc activates telomerase',[5] reports that when human lung tissue cells (IMR-90) are provoked with the telomerase catalytic sub-unit, hEST2, they do indeed top up their telomere quota—but despite that they still senesce and perish quite soon after reaching their usual Hayflick limit.

By contrast, other IMR-90 cells do get immortalised when forced to express a version of the Myc protein, which is expressed by a gene that causes cancer (an oncogene). 'We show that Myc can bypass replicative senescence under circumstances in which telomerase alone is ineffective,' note Jing Wang and colleagues. 'Thus, telomerase activity in tumors may simply reflect activation of oncogenes such as *Myc*', which also turn on quite a few other genes to do their dirty work for them. That is, observed telomerase production might be just one path into pre-cancerous activity, a marker but not a prime cause of unchecked life span. As Blackburn concludes: 'So we're back to not knowing whether expressing telomerase really *is* important in human life span.'

Cancer cells, alas, do already routinely perform this proliferative trick. They regain the vivacity of youth and burst into frantic, gobbling life. If we learn how to turn off *their* telomerase repair system, by introducing a tailored antagonist, maybe we can defeat tumours. This was one of Geron Corporation's pharmaceutical ambitions—to unlock a cure to cancer. By contrast, turning the repair system back on inside ordinary cells could make them—and us—immortal.

CONSTRUCTING A MOUSE

Several objections have been raised to this hopeful story. First, problems with telomere maintenance, by definition, can only afflict those somatic cells that make copies of themselves during an organism's lifetime. Plenty of cells don't—muscles and nerves, for example—yet they still age, along with the rest of the body. Is that just collateral damage, as the military might say? If the other, proliferative cells did not suffer from repeated and cumulative damage to their chromosome end-caps, might their nonreplicating neighbours have a better chance? Secondly, even those cells that proliferate and senesce often perish before their telomeres are gone. Maybe it's enough to doom a cell to have just one chromosome run down its stock of telomeres. But if that is so, what would happen, thirdly, to a creature without *any* telomerase, even in its stem cells and its germ-line cells? Would it peg out quickly? Surely it would not be able to bear healthy young, for they would be created from germ cells with abbreviated telomeres.

You could answer that last question by examining creatures that do not use telomerase, but that might evade the issue since they could have devised some alternative method of repairing telomeres. In this astonishing era of genetic engineering, a far more direct method is feasible: construct a test animal which does normally use telomerase but in which its gene has been identified and snipped out from the germ-line. Such experimental animals are routinely produced these days, and are known as 'knockouts'. Carol Greider and colleagues at several labs did indeed construct knockout mice lacking the relevant gene. The mice did not die young. And they had offspring.

Damn. So much for the telomere theory of ageing!

Well, perhaps not. Let's not be hasty. The species Greider used, *Mus musculus*, starts out with much longer telomeres than we possess, even though their life span is only three per cent of ours. Maybe they don't really need that degree of redundant TTAGGG coding. Maybe what looks like an excessive number of telomere units is a genetic error that hasn't been weeded out, unnecessary duplications akin to the junk DNA that litters mammalian chromosomes. It doesn't do any harm, but normally it doesn't do any good either. But in a strain deprived of the telomerase gene by knockout intervention, and hence unable to replenish its germ-line telomeres, this bonus might help the animals retain fertility and somatic integrity for several generations.

Eventually, though, even the redundant nest egg can be expected to run out, and disaster will fall upon that mouse lineage. So their children will have shortened telomeres, and their grandchildren shorter again, and so on, until crisis suddenly strikes. Then you'd expect to see various pathologies break out. After all, telomeres serve to keep chromosomes from sticking together, and organise the accuracy of their pairing up during division. Perhaps these ills would start to affect a later generation. Such an outcome wouldn't demonstrate a link between telomere maintenance and ageing, but it would provide evidence that telomerase is crucial in preserving the integrity of replicating cells. And that's just what happened to Greider's mutant strain of mice. The sixth generation, which lived as long as their parents, were sterile.

Different genes can serve to sustain telomere health in cells other than the germ-line, although the details are not yet understood. Maria A. Blasco, a principal investigator at Madrid's National Centre of Biotechnology, reported that 'the cells derived from these mice without telomerase have increased genomic instability (end-to-end fusions and aneuploidies [that is, too many or too few chromosomes created in the next generation, with injurious results]) as a consequence of telomere loss from chromosome ends.'[6]

At a recent US Medical Research Council Human Genetics symposium, Greider, Blasco and others reported that 'telomere length shortened progressively with each generation ... Analysis of litter size at each generation showed a progressive decline from the first through the sixth generation, and to date, no seventh generation of mouse has been produced. Histology of the testes in sixth generation [knockout] mice showed a lack of germ cells, although the support tissue appeared to be normal.' Indeed, cells tended to trigger their own apoptosis or 'suicide' genes if forced to proliferate. Greider noted that shortened telomeres in late-generation knockout animals might indicate the existence of a specific DNA damage pathway that induces apoptosis in cells so afflicted.

So enhancing the telomere repair system might, after all, help ensure that this sorry fate does not afflict our own ageing proliferative cells. Perhaps we are indeed on the road to physical immortality.

Again, I must stress that not all the experts are sanguine about the implications of such research for enhanced human longevity. After all, what goes wrong in ageing and what goes wrong in sixth-generation telomerase knockout mice have very nearly nothing in common. James Ryley, a young researcher working on characterising the differences in gene expression that occur during ageing, has briskly enumerated several reasons for doubting the centrality of telomere loss.[7] Most replicating tissues, such as skin, senesce well before their telomeres are exhausted. Many of the cells that fail with age, such as those in the heart, don't replicate anyway, so the state of their telomeres is presumably irrelevant. Yeast cells, in which telomerase is always active but which have no immune system, are just as mortal as humans, as are some mature insects which have *no* dividing cells. His conclusion: 'At our current life spans telomerase probably has little to do with anything save maybe arteriosclerosis ... Aging is almost certainly a multi-causal chain of events, with telomerase being a single link (and not the weakest link) in the chain.'

So scepticism is understandably more prevalent than optimism. For example, biodemographer S. Jay Olshansky of the Department of Medicine and Center on Aging at the University of Chicago has published a great deal of research into human ageing and prospects for longevity. In a 1998 article in *American Scientist*, 'Confronting the Boundaries of Human Longevity',[8] with biologists Bruce A. Carnes and Douglas Grahn, both of the Center for Mechanistic Biology, Argonne National Laboratory, Illinois, Olshansky argues that so many of us now live far beyond our reproductive years because the rugged engineering built by evolution into the species is bolstered, but only up to a point, by technology's protective environments. We are like race cars: not *designed* to fail, just not fashioned for extended operation.

This has its advantages. If ageing is due to evolutionary *neglect* rather than intent, 'there is every reason to be optimistic that the process is inherently modifiable'.[9] We might do this by altering crucial genes—an approach already under way—or indirectly by controlling the proteins that genes express. There might be an unexpected price. It could turn out that all we've done is shift the burden to different diseases and defects, or allow the appearance of genetic products and their consequences that today are masked by inevitable death. Olshansky and his colleagues dub effective bids the creation of *manufactured time*.

One hopeful sign of progress is a finding by Bruce Ames of the University of California, Berkeley, who has recently shown that accumulated damage to rat mitochondria can be reversed by pharmaceuticals. The Ames lab has reversed some of the functional decline in cells whose mitochondria are working poorly, although their DNA remains in good repair.[10]

Even so, modifying a human's longevity to any major extent would require such extensive and deep-seated changes that no simple, single adjustment could possibly do the trick. The recent telomerase studies, Olshansky told me, 'do not suggest that whole organisms can be immortalized. In fact, I seriously doubt that manipulating telomerase will even extend life at all in humans.'[11]

CURING (SOME) CANCERS

Such sensible caution is always welcome in such a speculative domain. By a neat coincidence—or so it might seem, except that all these discoveries are interconnected by the same new science—in May 1998 an equally astonishing claim had surfaced. Harvard University's Judah Folkman and his team at Children's Hospital in Boston showed that, in mice at least, two naturally occurring proteins called angiostatin and endostatin can selectively cut off the greedily growing blood supply tumours build for themselves while leaving normal tissues alone. This artful technique can shrink cancers with minimal side effects, since only the tumour's deranged angiogenesis program is interrupted.

Folkman's breakthrough buzzed through the Internet before most of the media heard of it, although it had first appeared in a *New York Times* front-page story by science reporter Gina Kolata. When I forwarded the news to my brother, a medical practitioner, his e-mail reply was vivid: 'This is *sensational!* My pulse quickened, and my eyes half popped out of my head. It reminds me of the first time that I read of the double helix ... The elegance and simplicity in combination hint that "this is right".'

Almost at once, cold water was thrown on such hopes, but not before understandable flurries of excitement. Stock in EntreMed Inc. (US), which holds drug rights to the two statin proteins, had exploded from a Friday close of $US12 a share to Monday's $US83. Kolata was offered a two million dollar advance for a book on the breakthrough. By week's end, though, the book deal was off, shares were down to $US30, James Watson and other molecular biology luminaries were backing away from their earlier reported raptures, and scientists in the oncology field stressed that this work was so far effective only on mice. Preparation of industrial quantities of the statins was untried and posed formidable problems. In general, the party died out with a whimper.

Still, the basic facts remained—the cancerous animals in the experiments had indeed recovered, their tumours starving to death as their blood supply cut off. That method might only apply to certain kinds of cancer—leukaemias, for instance, escape this counterattack—and we will need to wait for proper human applications. Still, if it does prove out in clinical trials, we're on the verge of mending at least some kinds of the second most frequent cause of death.[12] Equally cautious hope can be extended to the discovery of the gene *PTP-TD14* by Dr Mingdong Zhou and a research team at the Victor Chang Cardiac Research Institute in Sydney announced in August 1998 in the *Journal of Biological Chemistry*. The protein encoded by *PTP-TD14* informs cells of the proper contact boundaries between them, and appears to act to retard the wild growth of cancerous cells (technically, it 'may be critically involved in regulating Ha-*ras*-dependent cell growth'). It might provide a pinpoint therapy against the proliferation of abnormal cells, but clinical trials will take years to test this possibility.

Resolving all cases of cancer (a very difficult task, since there are many different ways in which tissues can become cancerous) would add an average three years to human life expectancy—and far more to those luckless individuals now doomed to early death by breast, bowel and other cancers.

THE MYRIAD WAYS WE DIE

Together, heart diseases, cancers and stroke kill close to three-quarters of those over 65. In the USA in 1991, more than two in five deaths were due to heart or blood vessel conditions; strangely, a far higher proportion of those were black people. Consider these figures from another typical western nation, Finland, for 1995: cardiovascular ill-nesses accounted for a little under half of all deaths; cancers, just over one-fifth; respiratory disorders and gastrointestinal disease, a bit more than a tenth. Nonmedical causes (unless one includes psychiatric motives), including accident and violence, about nine per hundred; suicide, about three per hundred. These figures are not very different from Australia in 1991, where cancer caused a little over a quarter of all deaths (while suicide was listed on about one per cent fewer death certificates, perhaps for social reasons). In 1991, the USA ranked sev-enteenth of 35 nations surveyed for cardiovascular death. The highest heart death rates were in the former Soviet Union, Romania, Poland, Bulgaria, Hungary and Czechoslovakia; the lowest were in Japan, France, Spain, Switzerland and Canada.

Solving cardiovascular diseases would add 14 extra years to the popu-lation average—a truly sensational benefit to the old. But that might take more than cleverly engineered drugs. Ultimately, swarms of nano-robots the size of viruses might have to be used to scour our blood vessels, cleaning out plaque and even surgically repairing damaged tissues. (We'll return to this extravagant vision in the next chapter.) As well, using knowledge gleaned from the ongoing Human Genome Project, we'll tweak our recalcitrant genes into recovering their self-repairing abilities.

All that might sound like pie-in-the-sky, but so did a cure for tumor-ous cancers even in mice, not many months ago. Results pouring from research labs strongly suggest that we're accelerating into an almost unimaginable future, with unprecedented hazards and benefits. More detailed understanding will emerge in those labs during the next decade or so. Subtle control of telomere maintenance will heal specific damaged tissues, and help extend human life spans.

There's more to ageing than telomeres and cancer, of course. Cells are damaged by stray radiation, toxic chemicals, genetic copying blunders and other hazards. Eventually we will be able to proofread these errors and correct them. Dedicated molecular machinery, unused in most of the body's cells but coded in their core DNA, already protects sperm and eggs to a significant extent. (Old patriarchs do not sire ancient children, although perhaps their children have an enhanced chance of genetic defect, which is certainly true of the offspring of mothers nearing menopause.) This proofreading and damage-containment machinery might be craftily reawakened in the rest of our cells. That, at least, is the kind of argument standing behind my claim that the last mortal generation is now on the earth—and perhaps the first of those who will never need to die.

THE BREATH OF LIFE AND DEATH

Possibly the worst cellular injuries are caused by *free radicals*, ionised chemicals that wander through the body locking onto proteins, DNA, lipids, whatever will bond with them. A free radical is just a molecule which has one or more unpaired electrons. (Relax—there won't be a graded test at the end of this book, so do feel free to skim the next few pages. But it's really not that frightening, and a sense of these basics will help you understand why we age and what we might soon learn to do about it.)

Free radicals are energetically most stable when they lock into combination with other structures, but the new compounds then acquire different properties, and generally can't do their jobs any longer. Evolution has built countervailing systems to mop up as many free radicals as possible, but they are endlessly replenished by the very processes of metabolism that keep us alive and on the move.

Deprived of oxygen, we gasp, flail desperately, and soon die. It is the fuel of life, making up 23 per cent of the air, 86 per cent of the oceans (where it is locked up with hydrogen in the form of water) and 47 per cent of the solid crust. Paradoxically, it is also a ferocious chemical and biochemical poison. Slumped, pitted car hulks abandoned beside the highway, corroded by brown rust, are the victims of oxygen. So are we. Metabolism, the slow burning of our food, is an oxygen-mediated process, but so is the deterioration of many crucial tissues. Francis Crick, co-discoverer of DNA's helical structure, put it this way:

> The great advantage of oxygen is that it permits a cell to obtain far more energy from metabolizing its food ... Molecular oxygen is a powerful but dangerous compound. It is potentially a highly toxic substance for cells, because cellular processes are liable to produce several lethal derivatives of it, such as hydrogen peroxide (H_2O_2) or an even more dangerous compound, the free-radical superoxide (O_2-).[13]

Can't live with it, can't live without it.

Life uses oxygen as an energy lever. Muscles, for example, turn sugars into lactic acid in a fermentation process that releases usable energy. They recover after a stint of work (again simplifying to a disgraceful extent) when stored lactic acid is broken back down into more primitive waste components—carbon dioxide and water—by oxygen atoms, which usually travel in somewhat stabilised pairs, O_2. (It has recently been claimed that chronic fatigue syndrome—mocked for years as 'yuppie flu', pilloried as mere malingering—correlates with a disorder of the lactic acid removal system. It seems this fatiguing substance accumulates to drench the weary muscle tissues. As always, though, everything is dynamic balance, for lactic acid is what's known as a *chelating agent*, an ion scavenger, and in proper amounts helps extend longevity.) *Oxidation* occurs when carbon, say, is 'burned'—combined with two energetically cosy oxygen atoms to form carbon dioxide (CO_2) —or when iron rusts by adding three sets of molecular oxygen ($3O_2$) to four iron atoms ($4Fe$) to yield $2Fe_2O_3$.

A single oxygen atom readily bonds to a pair of hydrogen atoms, forming an H_2O water molecule. When the four available slots of a single carbon atom are filled by three hydrogen atoms plus an oxygen, the remaining spare oxygen bond can couple to a fourth hydrogen—resulting in the simplest alcohol molecule, methyl alcohol.

In cases like those, the oxygen molecules are safely kept out of circulation. It need not be that way. Oxidation is a slightly sloppy term, applying not just to cases where oxygen locks onto another molecule but also where hydrogen is filched from a substance or it loses one or more electrons. Hence the minus sign in the superoxide symbol (O_2-) in that Crick quotation; it's shorthand for a spare negatively charged electron. These are the prescriptions for rampaging free radicals with unpaired electrons.

During metabolism, many free-floating and highly reactive oxygen radicals get released into the delicate economy of the body's tissues. Unless they are adroitly mopped up, free radicals are attracted to other compounds by the energy imbalance of their unpaired electrons, adhering to their hosts and, in the process, frequently damaging the tissues of which they form part.

So the reactive oxygen species (or ROS) varieties of free radicals are a major source of trouble. They can damage both the cell's nucleus, where the DNA resides, and the small factory and energy organelles of the outer cytoplasm. Each cell's energy is stored and released by mitochondria floating in the cytoplasm, now thought to be the descendants of bacteria conscripted inside more sophisticated cells as life became elaborated billions of years ago. Mitochondria have their own loop of DNA, coding for thirteen proteins, inherited only from the maternal line. These luckless relics are less efficiently protected by the more up-to-date enzyme systems that nuclear DNA uses to repair oxidative damage.

More than 40 years ago, Denham Harman proposed that ageing is principally an accumulation of damage to key parts of the body's cells wrought by free radicals. In 1972 he suggested that mitochondria might be most at risk, since these chemical powerhouses are a prime source of oxygen free radicals. Certainly some mitochondria fall prey to ROS impacts that delete or rewrite genes, compromising the specialised function of the cell. But it turns out that less than one-tenth of one per cent of mitochondria in mammalian tissues show such damage. Could such a modest effect interfere drastically enough with the cell's—indeed, the body's—metabolism to bring about the tragic cascade of ruin we call ageing? Despite Harman's advocacy, and research continuing through the decades, it seemed that the cellular mechanism responsible for ageing must be found elsewhere—in telomere shortening, in changed patterns of gene expression and so forth.

In the first issue of the *Journal of Anti-Aging Medicine* (1998) its editor declared that this estimate was due for review. The mitochondrial camp, the editor observed, had 'lacked any coherent rationale for the preferential accumulation of damaged mitochondria in aging; the battle appeared to have been won by the senescent gene expression camp not for reasons of data, but rather by intellectual default'. In one stroke, however, in February 1997, a paper by Cambridge University theorist Aubrey D. N. J. de Grey had shifted opinion sharply, providing an elegant mitochondrial explanation for ageing, and suggesting bold implications for clinical treatment.[14]

De Grey noted that even cells which do not divide—muscles and nerves, for example—nonetheless produce a continuing turnover in the generations of their mitochondria. Why do cells bother with the energy-expensive business of breaking down old mitochondria and making new ones? De Grey offered a startlingly counter-intuitive explanation: the housekeeping machinery of the cell *preferred*, by a kind of bungle, to destroy healthy mitochondria while leaving the mutant, damaged units alone—so that some cells ended up choked with poisoned, useless mitochondria. How could this be explained in a system that had evolved by natural selection? Isn't it supposed to be 'survival of the *fittest*'?

The argument is enticing, however. Suppose everything is functioning happily, and mitochondria are doing their job of providing ATP, the chemical that powers cells. This process releases a lot of reactive oxygen species, which attack the membranes of the mitochondria. 'In due course,' de Grey argued in his original paper, 'the membrane will become unable to perform its main function, which is the maintenance of the proton gradient created by the respiratory chain. This will not affect the operation of the respiratory chain itself, however, only the production of ATP, so damage will continue.' A suitable strategy to ensure a cell's supply of ATP is to destroy mitochondria that have become damaged in this way, and to prompt the remaining less-damaged mitochondria to divide, thereby maintaining the number of mitochondria in the cell as a whole. But suppose a mutation strikes that impairs a mitochondrion's respiration, lowering its production of harmful free radicals. Suddenly the *mutant* organelles are damaging themselves with their own free radical production *less* than healthy ones are—so they survive better than their optimised kin, which are being digested by mistake!

In rapidly dividing cells, this is not a problem, since membrane damage is cancelled out by fresh membrane creation. Brain cells and muscle tissue, alas, do not replace themselves—although their mitochondria do, providing opportunities for mutation and 'an accumulation of ATP-deficient cells leading to aging at the organismal level'. If de Grey's argument is correct, mutated mtDNA can thus be very damaging even though there is so little of it in the body. Cells with mutant mtDNA seem to survive indefinitely, despite having lost the ability to use oxygen. These cells may be highly toxic, releasing ROS molecules into the bloodstream and thence into other, mitochondrially healthy, cells. It is this increased load of *oxidative stress* that wrecks lipids and proteins, with those ill effects multiplied to other, healthier tissues: a classic example of throwing a single spanner into a complicated machine and watching all the gears seize up.

ROS molecules also attack DNA in the nucleus, of course. Bruce Ames of the University of California, Berkeley, estimates that the nuclear DNA in every cell is struck by ROS ten thousand times every single day. Fortunately, maintenance systems usually manage to detect and snip out most of these ruined segments, and rewrite the code correctly. Still, this is an assault on both the memory of the cell and its ability to control and organise the tasks allocated to it by its specialised role. Earl Stadtman of the US National Heart, Lung and Blood Institute concluded that half the proteins in the bodies of the elderly might be damaged by oxidative impacts. Many of those proteins would be key enzymes, needed to keep the body working effectively. Clearly, methods to reduce or recover from the assaults of ROS radicals would seem a good path to extended, healthy life.

SUPEROXIDE DISMUTASE

Why doesn't the body do something to mop up ROS agents? It does. A whole armamentarium of antioxidant substances is either made by the healthy body or imported by it in food. Vitamins E and C are well-known examples. Others include glutathione peroxidases, catalases and uric acid. There are dedicated proteins such as heme oxygenase and one of the most important, *superoxide dismutase*. Just to give you the slightest tang of how complex this field of knowledge really is, here's a brushstroke portrait of SOD (infelicitous abbreviation though it is).

The enzyme superoxide dismutase is a dimeric or two-part protein with a zinc ion and a copper ion in its active sites. It's the positively charged copper ion, in a hollow at the base of the protein, that does the chemical trick. SOD, like all enzymes, is a kind of introduction service for other molecules, or sometimes more like a divorce lawyer. It isn't changed itself in the process, but it serves to remove the destructive superoxide anion radical. (An anion is just a substance with a spare electron.) It swiftly dismantles or 'dismutates' O_2- into molecular oxygen and hydrogen peroxide. However, hydrogen peroxide is itself heinously dangerous to cells and their membranes, and the enzyme catalase specialises in nabbing it in turn and breaking it back to benign molecular oxygen and water. The chemical paths whereby SOD and catalase carry off these small feats are complicated and tediously incomprehensible to anyone except specialists. Fortunately, we don't need to follow the explanation all the way down to gain a clear sense of what's at stake.

When the normal gene that produced SOD, *SOD1*, goes on the blink after its structure is altered in a mutation, one consequence is motor neuron disease, or amyotrophic lateral sclerosis, the progressively crippling disorder that has kept Stephen Hawking in a computerised wheelchair for many years. In March 1993, researchers led by Teepu Siddiqu at Northwestern University, Chicago, showed not only that the disease appeared to run in families, often the mark of a genetic cause, but localised it to a faulty *SOD1* gene that is preferentially expressed in motor neurons. According to a June 1998 paper in *Nature Genetics*, reporting an experiment on fruit flies, 'overexpression of a single gene, *SOD1*, in a single cell type, the motorneuron, extends normal life span by up to 40% and rescues the life span of a short-lived *SOD-null* mutant.'[15] The search for a cure in humans continues.

At the bottom line, the question is this: if we could force our cells to churn out more of the properly formed antioxidant defenders, or altered our nutrition to increase the amount we process through digestion, could we retard the ageing process at the cellular level? It seems so. In 1988, Thomas Johnson of the University of Colorado showed that knocking out the gene *age-1*, by mutating it, boosted the life expectancy of the worm *Caenorhabditis elegans* by 70 per cent.[16] How so? In the absence of the protein products of that gene, *C. elegans* cleared away an unusual amount of free radicals by somehow producing extra quantities of superoxide dismutase and the enzyme catalase (the opposite effect, in a way, to the evil consequences of a *SOD1* mutation). We'll come back later to the bizarre fact that some standard genes *impair* the longevity of certain creatures (which implies that *at a more youthful point in their life cycles* those genes do more *good* than harm—which you pay for later by dying!).

What's more, in 1994, research scientists W. C. Orr and R. S. Sohal reported in *Science* that 'upregulating' or overexpressing superoxide dismutase and catalase genes in fruit flies, an animal more complex than *C. elegans*, increased their life spans by a third and more.[17] These genetically modified animals ended up with less damage at both the molecular and organic levels. Other ageing-related genes had already been located by Michael Jazwinski, of the Louisiana State University Medical Center. The mnemonically named *LAG1* (short for *Longevity Assurance Gene 1*) was found in brewer's yeast, a creature even more rudimentary than a worm, which multiplies by budding. When extra *LAG1* product is expressed in old cells, their yeasty life span is extended by a third, while they happily remain non-cancerous. Best of all, a version of one yeast longevity promoter is also found in human tissue, suggesting that this might work for us, too.

At the very least, it isn't obvious that similar modifications would fail to increase human life expectancy by an equivalent amount. On the other hand, as genome researcher L. Stephen Coles cautiously noted in the *Journal of Longevity Research*, 'There are enough idiosyncratic differences in the metabolic physiology of insects to cast suspicion on any attempt to claim more than what was observed with the species in question.'[18]

Even if inserting genes for anti-free radical agents does work successfully with mammals, perhaps it would need to be done at the embryo stage, unless cunningly wrought retroviruses can be contrived to port the appropriate genes into all the cells of our adult bodies. Some might find this an absurd or horrifying suggestion. Tinker with your baby when it's an embryo? Still, it might prove an unavoidable path if we are to take advantage of several longevity prospects that are already known to work in other species—if only, so far, in the lab.

The new approach suggested by Aubrey de Grey hints at means to improve mitochondrial health, and thus repair the energy systems of the body. De Grey has offered several suggestions, some more extreme than others and all of them too technical to discuss here in detail. One path would maintain or restore oxidative phosphorylation, which accounts for nine-tenths of synthesised ATP. How? Using a suitable vector, engineered or normal 'wild-type' mtDNA could be *imported* into ailing cells, to augment damaged genes. Or residual wild-type mtDNA (if there is any left in a given cell) could be encouraged to replicate. Or the thirteen protein-coding genes that the organelle carries could be reverse-engineered into the nuclear DNA itself, using tailored transgenes. Protein products expressed by normal mtDNA (now safeguarded in the nucleus by its superior proofreading machinery and its lowered incidence of ROS damage) could be built in the cell's cytoplasm and moved through the mitochondrial membrane. This would also require new apparatuses to force the expression of needed proteins, chaperone them into the mitochondria, and fold them up neatly after they have unkinked for their passage through the pores of the outer and inner membranes. There are formidable obstacles in the way of each of these bold proposals, but de Grey argues that at least one method—importing proteins—can perhaps be solved, possibly by learning from the equivalent genes in plants (which employ the 'universal' nuclear DNA code, instead of the variant version used by animal mtDNA).

But why not just eliminate the guilty party, the less than one per cent of low-efficiency or anaerobic mitochondria? If de Grey's analysis is right, cells harbouring these organelles have certain distinctive features—technically, a very high level of plasma membrane oxidoreductase, or PMOR—that could serve as a well-defined target. What's more, a reagent called pCMBS selectively inhibits PMOR and thus kills cells relying on it. If the progressive enfeeblement of mitochondrial defences against antioxidant attack can be retarded or halted by such means, then very plausibly, as de Grey claims, it 'would profoundly alter our view of the inevitability of old age'.

Genes for Death

So far our emphasis has been on accidents and system failures. Wear and tear is inevitable, and even the best backup procedures are far from perfect. Evolution, as we'll see, has no stake (to speak metaphorically) in keeping us alive and healthy past a certain age, because human females stop reproducing in their late forties or early fifties. Menopause, as we've seen, is either a coded and systematic shutdown of the female body's reproductive cycle when her eggs are running low, or a messy blend of failures on a number of hormonal fronts. It looks to me at least somewhat designed (in the sense that anything produced by random variation and competitive selection can be so described), if only because we don't find at least some women of 90 or 100 still in the bloom of late motherhood. If menopause were just a statistical average of various deficiencies in the body's maintenance programs, that prospect might not be quite as absurd as it sounds. As promised, we'll return to this topic in the third chapter, when we look more carefully at the Darwinian pressures that encourage ageing and death in individuals.

We do know, though, that humans and other creatures possess some nuclear codes that quite explicitly and fairly can be termed 'death genes'. These do not necessarily have any directly deleterious impact on the body as a whole—in fact, they are crucial to making us who we are, and can save us from cancers—but they are precisely targeted at the death of individual cells. The process they trigger is called apoptosis, or cellular suicide.

That's essential in early embryonic growth, and during early learning, because the number of explosively multiplying cells of the new-made body wildly exceed requirements. Development, as I noted earlier, is a matter of pruning and sifting. What's more, the immune system needs fine-tuning to ensure that its killer cells don't turn ferociously upon other somatic cells bearing the body's own genetic signature. We churn out a vast diversity of differently coated immune cells, a kind of pre-emptive array of defences against every conceivable antigen foe, and these roam the body waiting to run into an intruder. (This kind of military metaphor offends some people, but it is hard to avoid when what is at stake is a life-and-death conflict—admittedly at the level of mindless cells.)

Once a single immune cell locates an alien cell it doesn't approve of, it marks it and starts to clone itself, ready to engage the enemy using its idiosyncratic specialised weaponry. A defensive system of this sort inevitably finds some of its mindless soldiery poised to assail its own cellular signature, and that can't be permitted. (When it does happen, we get awful autoimmune disorders that are very resistant to treatment.) Usually, then, certain self-defensive signals tell these treasonous soldiers to lay down their arms and die—and they do. Nuclear DNA splits into useless chunks, membranes collapse, the suicidal cell is digested and dissolved.

Apoptosis can also be triggered in cancerous cells, ordinary body tissues that have gone out of control and started to proliferate without ever reaching their mature form. These are set in motion by proto-oncogenes that trigger wild over-expression of a given protein. Many pharmaceutical researchers are now looking hard for substances that will reliably set off cellular suicide in tumours marked by such errors, while leaving healthy cells alone. Certain proteins have been identified that suppress cancerous tumours by forcing them to commit suicide. The most well known is p53. Encoded by the gene *p53*, it recognises when a cell's DNA program is badly damaged and acts to block cell growth and thus kill the deviant cell from within. In response, some cancers and viruses (such as the common cold) have learned to switch off the codes that launch manufacture of the p53 protein. In fact, faulty *p53* genes are found in about half of all cancers.

The gene *BRCA1*, isolated in 1994 on chromosome 17 by Mark Skolnick and colleagues at Myriad Genetics, Utah, is somehow involved in protecting against breast and ovarian cancer by assisting p53 protein in its apoptotic work. It's called a 'coactivator', and works to regulate the transcription of DNA into RNA, the messenger which totes a copy of the gene's instructions to the cell's protein assembly factories. When *BRCA1* takes a faulty form that fails to help kill tumours, the person carrying the gene is at greatly enhanced risk. There are, however, many ways for *BRCA1* to mutate, and not all of them have lethal consequences. Chromosome 13 contains another breast cancer-related gene, *BRCA2*, with defective forms having similar dire effects.

The question arises, then: do *all* our cells, our organs, our entire living system, have specialised apoptotic genes that kick in as we approach the telomeric Hayflick limit, or when sufficient of our body cells have been damaged by oxidants, raw sunlight, physical abrasion and other mutagens? Are there literally dedicated 'death genes' for the whole system that could be isolated and modified—switched off entirely, perhaps?

I think the evidence at present stands against that hypothesis. We endure, probably, for as long as our cellular integrity is well maintained. That can vary among different kinds of cells, and from one organ or system to another, as we've seen. Parts of the brain continue to work quite well even for centenarians, as do most muscles (however weak they become by contrast to youthful vigour). Why, then, do we lose the maintenance systems that kept us hale when we were in our teens and twenties? Why doesn't evolution hit upon better antioxidant defences, or use less corrosive materials in the first place?

It seems likely that the answer is a heartless matter of statistics. We are what we are because our parents were quite similar to us, and they survived long enough to conceive and rear us. Once they provided two or more healthy, fertile offspring, it wouldn't have mattered to the statistics of species survival if they'd perished at once. Many animals do just that. They live for a day or a few years, bear young, and die long before they ever reach their Hayflick cell limit or grow old.

That explanation raises some curious puzzles of its own, you'll notice. It still doesn't explain *why* parents die, since clearly if they remained healthy and vital they could keep adding to the population pool of their species just as their own children will—more so, perhaps, since many embryos are lost long before birth, some babies are sickly and never succeed in breeding, whereas their parents are proven survivors. Again, we'll come back to this vexed topic.

For now, we need to notice that a genetic effect the biologists call *antagonistic pleiotropy*[19] is likely to harm the elderly without even meaning to. Suppose you start with an effectively immortal population of grils, an imaginary animal I just invented. This is actually how life worked for a very long time, when every living creature reproduced by fission— although, as we've seen, 'immortal' might be a misleading way to regard sexless organisms, which don't so much persist as echo themselves identically in their cloned offspring. Suppose, further, that a random high-energy particle from space or the bowels of the earth smashes through the germ-line DNA of one of these grils, and alters the protein expression of one of its genes.

It turns out that when the offspring (or phenotype) carrying that mutation (or genotype) is born and matures, it produces twice as much hair on its body. As it happens, the winter of the world is setting in, a new ice age inexorably cooling the planet. If temperature had been going the other way, this gril and its descendants might have fared badly, stifled by its shaggy coat. Luckily, it thrives in the cool evenings, mating successfully with its sparser-haired companions. Although grils are immortal, in the sense that they don't age as we do, that doesn't save them from infestation by parasites, catching diseases and dying of illness, being eaten by predators, or dying of exposure in the ice age chill. Over the centuries, hairy grils cover the plains.

Alas, there turns out to be a hidden biochemical cost to that handsome pelt. The gene or gene complex that determines the shagginess of the gril coat does so, as it happens (it need not have been this way), by expressing an enhanced level of putradine, an unpleasant chemical that accumulates in the gril liver and reaches toxic levels by the time the animal has produced, on average, three or four litters. Death of a new kind enters the gril world—mortality as we know it. It is the price paid for pleiotropy, that is, many alternative forms of a given gene (in this case, for hirsuteness).

Clever equations exist for probing this explanation, curves that rise and cross on a graph. At a certain point, there will always be some stage at which a potentially immortal creature producing offspring regularly, but at risk of death by natural causes other than senescence, will be outperformed by a superior sibling with a fatal mutation that only kills after reproduction. 'Superior' in this sense has a grim irony to it, admittedly, since it doesn't help an individual to be stronger or smarter or more fertile if he or she is being poisoned by the very gene products that contribute to youthful vigour. But natural selection—or rather, competitive retention—doesn't care. Nature is mindless as well as heartless, a point we'll return to again and again. Nothing counts, finally, in the shaping of future generations except numerical superiority. If you attain that for your descendants by dying young, so be it. All those parents who grumble from time to time that their kids are killing them are right—evolutionarily speaking.

This fanciful story about grils and putradine is purely imaginary, of course, but the principle is solid. A real-life example of antagonistic pleiotropy in humans might be the link between levels of circulating oestrogen and breast cancer. Fertility depends on this hormone, but it aggravates tumours that chance to start growing in the breast. That doesn't mean the link is direct. It is not. Recall the two genes, *BRCA1* and *BRCA2*, that in abnormal forms can launch the sequence leading to cancer of the breast. If such defective genes were corrected, perhaps it wouldn't matter that a woman's body is flooded with oestrogen from menarche to menopause. As it is, though, there is a tragic link. Having children the mammalian way puts a potential mother at risk of death. Sex and death, as artists have claimed for centuries, are closely allied.

But that need not be the end of the story. Knowledge can alter fate, in both art and life. We know more now about these links, and we are starting to grasp the deep, subtle and often loathsome tangles of cause and effect in life and death. With more work, it is not absurd to hope that within a generation or two we might have uncovered enough to slow, halt, or even reverse ageing, and return ourselves to the primeval bliss of the early grils.

THE 'PLAYING GOD' BOGYMAN

But wouldn't that be ... *unnatural*? Isn't the conquest of ageing and death an affront to the divine Creator? This objection tends to get wheeled out every time something unexpected heaves into view. Let's put it to

the test of recent history and the way our views have grown sophisticated in the face of unforeseen opportunities and benefits. Does God
forbid the use of analgesia during childbirth? That cruel opinion was
brandished in the suffering faces of women for generations, before it
was scorned into silence by the determination of mothers not to remain
in pointless pain just because of a craftily interpreted verse in an ancient
scripture. Is there a divine rule against adjusting aged eyes by popping
on a pair of reading spectacles or contacts, or having the eyes' lenses
reshaped by laser surgery, wondrous boons not known to the ancients?
I certainly hope not! And even if there is, I doubt that many people
would obey it.

Still, doesn't the accumulated wisdom of humankind tell us to beware
of meddling with the deep secrets of life? Consider this thought-
experiment:

The aliens land in front of Luna Park or Coney Island (did you really
think they'd choose the White House lawn?), and they're a thousand
years ahead of us, or a million. So they already know it all: the Theory
of Everything, the scientific truth about ... you know ... God and the
purpose of life.

Do we handle the news well?

We do not.

It turns out that both God and the divine purpose are entirely
random, for all practical purposes, utterly unconcerned with our happiness. God's designs alternate, the aliens tell us, between the heedless
and the sickeningly cruel.

Luckily, we needn't wait for aliens to bring us that shocking news.
We already know that *is* actually how things are. Charles Darwin learned
as much a century and a half ago, and the disturbing truth is still filtering
down, more clearly with every decade of evolutionary research, beyond
doubting, remorseless. I cheated in my thought-experiment, of course,
by using the word 'God', which has grown specialised and ambitious
since Yahweh's supporters put all the wild gods to death. To the extent
that the universe has a 'plan', it is this: *the multiplication of coded, slowly
mutating replicators, Richard Dawkins' 'selfish genes', carried down the stream of
time in disposable bodies.*

The meaning of life is more life—but not your life or mine, or the children's.

Most people really don't wish to hear this. Some of them are probably waving fatwas at the news, or igniting flaming crosses. I don't relish the news any more than you do. It would be comforting, in a way, to live in a world demonstrably designed for our special convenience, or even in an inconvenient world where we are the key players. On the other hand, it is arguably much better that we do not live in a universe where anything can happen at the caprice of supernatural beings. In any event, our preferences are irrelevant. Planless evolution *is* the way things are, here in the real, open-eyed, adult world. If we are to have the slightest chance of arranging human affairs more equably and profitably, we must start by learning how we got here (by natural selection in a planless universe) and how much latitude we can scrounge for ourselves in changing evolution's bleak dictates.

It's true that natural selection has not tumbled by itself to the telomerase rejuvenating trick, or superoxide dismutase enhancements, or any other ways to prevent bodies and minds growing decrepit. That's because evolution, alas, readily discards bodies and the minds they house, getting more reliable mileage out of starting afresh with a new generation. It's not impossible to design bodies for longevity, just less cost-effective, for the genes, in evolutionary terms. Evolution combs our genes, not us as individual persons. We have every motive to set its damnable methods right.

Why should evolution fail us so cruelly? Because bodies that stress efficient self-maintenance tend in the long run, like the thin-haired grils, to leave fewer offspring than those which ignore cellular errors once the initial breeding season is done. According to the disposable soma theory, the major determinant of a species' rate of ageing is its typical rate of death from causes *other* than ageing. If you can fly, or are brawny or brainy, you will be able to live a long time and choose a good summer (when there's lots of food around) to raise offspring. If you're low in the food chain, by contrast, you have no choice but to get on with breeding as quickly as you can. The former strategy requires slow ageing, a trade-off in which extra energy needed for maintenance is unavailable for early reproduction. Yet soon we will be in a position to *fix* the consequences of such blind evolution decisions, *from the outside*, once we know how to do it. We need not fear that by taking those steps we would be 'breaking Nature's law' or 'interfering with Evolution's plan'.

Despite the beautiful patterns of life, evolution *has* no plan. It is a gigantic, stupid lottery. Knowing that we dwell in such a universe, we are freed from fears of impiety (although not, needless to say, from the obligation to be careful, provident in our experiments and ventures). Since evolution does not have a plan for us, we may choose one for ourselves. In fact, that's what we have always done, whether we knew it or not. We have had no other choice. All too often we have hidden that deep moral responsibility from ourselves, delegating it to deities or 'Nature'.

Many people apparently find it hard to grasp this essential moral point. The day after the immensely important Wright-Shay telomerase discovery was published, a whimsical editorial appeared in my metropolitan newspaper poking cautionary fun at the 'tireless scientists'. The editorialist implied 'tiresome' as well, I suspect, for these scientific wretches were said to be pandering 'to this society's adoration of youth'. On the contrary—those benefiting most from healthy extended life span will be the elderly, especially if rejuvenation is part of the bargain. (That might be a comic threat the editorial missed: rule forever by deathless Baby-Boomers, or worse yet, immortal Gen-Xers!)

The commentary noted, as if this were a sound argument, that in Jonathan Swift's *Gulliver's Travels*, the immortal Struldbruggs 'were the most miserable of mankind'. Anyone who has studied the writings of Dean Swift knows him for a dyspeptic misanthrope. Why should we trust the opinions of such a life-defeated fellow?

One-sided fables like the Struldbruggs often get trotted out as if they prove something, as if they are *reports* rather than *inventions*. Frankenstein's hubristic fate is another, as is the Greek legend of Tithonus who lived forever, but in senility. All they prove, of course, is that people find vindication in their misery, not merely by reconciling themselves to the inevitable, but by embracing and even glorifying it. The barbarous practice of clitoridectomy—female genital mutilation, which surgically removes a girl's ability ever to experience sexual pleasure—is a ritual held in the custody of other women who have suffered the same fate, enforced not only by cruel male oppressors but by older victims of the same crime.

Death itself is more terrible than any 'punishment worse than death', because it is so *final*, and only acceptable because until now there has been nothing we could do to stave it off. This remark might sound callous to those in extreme physical or emotional pain, for whom death might seem deliverance from torment. Death can indeed be the sole available release from intractable pain, terminally enfeebled age, or devastating misery—and I believe everyone has the right to choose death if they so wish. Still, we stand now at a very special time in the history of the species. If we are truly the last mortal generation, perhaps in sight of the chance of becoming the first of the immortals, it surely would be best for us to refuse to bow in the face of ordinary fears and habitual scorn.

Defeating death and planning rejuvenation are goals no more absurd than finding remedies for short-sightedness or asthma. (I have been taking daily prophylactic drugs against asthma for more than a quarter of a century, and it has improved my life beyond recognition.) Because we are intelligent, resourceful, imaginative and brave, we manage quite well without an evolved ability to read and write at birth, or to fly a jet by instinct. Instead of hard-wired instinctive patterns, we use our brains, and our collective memory, which is why humans reached the moon while moths merely fly into the candle-flame.

This is not to assert that radical life extension will be a boon gained easily, or simply implemented. At first, it will doubtless take a multi-factored barrage of treatments, starting as it has done already with medical advice on sound nutrition, restraint in use of intoxicants, avoidance of lethal habits such as smoking, regular moderate exercise, vitamin supplementation, avoidance of excess physical and mental stress wherever possible, modest calorie reduction, perhaps some meditation. It is even possible that in the short term, something akin to the Tithonus myth *might* be the only available life-extending option. But who in their right mind would opt for the dubious boon of Tithonus? Some might.

Chris Lawson, a medical doctor and writer, comments on that possibility:

> *Most of us would hate to live this way, and would likely commit suicide once we had become too decrepit to enjoy life. However, there would be people with philosophical objections to suicide. There will be others who live in a society that objects to suicide/euthanasia and who lack the physical means to suicide because they are too decrepit. And these people, voluntarily or not, will adopt the Tithonus option.*[20]

But does the ever-more-frail option even make mechanical sense? Perhaps ageing in individual cells can be halted or reversed. Even so, Lawson notes, 'they would still be subject to the ravages of oxidation, mutation, etc., and on a macro-scale some of the damage might be irreversible, such as osteoarthritis or cerebral infarction. Our death-dealing genes may have been switched off, but our bodies will still decay through wear-and-tear and fat-rich diets and alcohol and breathing city air and cigarettes and sunlight. We will have escaped our genetic jailers only to find ourselves in a prison made of entropy. Unless nanotechnology can deal with this, we'll end up with the situation of a therapy to stop aging (in one sense) while allowing the tissues and the organism as a whole to continue to deteriorate. The Tithonus scenario is certainly possible (if unpalatable).'

Lawson adds, though, that it is improbable. Why? Because the death rate of a population of Tithonuses increases exponentially with age. As their bodies suffer the remorseless impact of the world, they have ever more chances of perishing by accident, sheer micro-level wear and tear, and novel diseases. Surprisingly, Lawson's mathematical analysis suggests a maximal life span for Tithonus of only 115 years. (Others, as we'll see in the final chapter, reach a more optimistic but no less final conclusion.)

So for effective life-extension, we will need to have perfect proofreading, repair and maintenance firing in all tissues, even if that calls for computerised nanomachines to suffuse the body. Such improvements are by no means improbable. Extra techniques would soon become available:

safe hormone replacements and boosters, effective antioxidants, improved inoculations against common and uncommon viruses and bacteria. Not long after that, it may be feasible to modify your individual genome to enhance the genes luck has dealt you, mask or edit out injurious genes, even add entirely new human-constructed chromosomes able to express novel and useful proteins (artificial chromosomes have already been made and added successfully to living genomes).

In the long run, I believe we will master our anxiety or revulsion over tinkering with the germ line. Corrections and improvements will be introduced into the genomes of the newly conceived. Those babies will grow up with bodies that need never die of old age. Will they still be human? I don't know. Perhaps they will be transhuman, or posthuman. Certainly, by the standards of our Pleistocene ancestors we ourselves are a startling new creature they might not have recognised as kindred.

IMMORTALITY BY DESIGN

We might reasonably look forward to some kind of anti-agathic treatment, to borrow the name the brilliant imaginative writer James Blish coined for such measures many decades ago. The prospects of immortality remain elusive, even so, because the best medical intervention in the world cannot unscramble an omelette back into an egg in its shell. There will be no simple B-grade science-fiction 'serum' or 'DNA transfusion' that will reverse ageing the moment we take it, almost instantly plumping out our cheeks and turning our white hair dark (or replacing it, in my case).

The reason why we need to keep a grip on our anticipations, even when demanding the formerly impossible, has been spelled out charmingly by Susan and Robert Jenkins in their book *Life Signs*. Lightly, but with considerable rigour, they critique the biological sciences of the 23rd and 24th centuries as depicted by the scientifically illiterate scriptwriters of the *Star Trek* mythos. Despite such media images, we're unlikely to run biological time backwards overnight. 'Some parts of us— the cartilage in our ears and nose, for example—keep expanding all our lives,' they note. 'Young beauties sport small, cute button noses. Old hags have long hooked noses.'[21]

Even those examples are more benign than the true horrors—the tangled, plaque-ruined brains of Alzheimer victims, the smashed immune systems of people living with AIDS, the general debility that makes old age a time of mental and physical frustration (however much tranquillity and peace can be gained by reconciling oneself to this time of decline). If the ears and noses of old immortals do keep growing, perhaps the world will slowly choose to see those characteristics as pleasantly neutral and even find beauty in them. Alternatively, it might turn out to be comparatively simple to tailor inhibitors, suppressors and promoters that can be applied to different tissues. If worst comes to worst, many millions of people already turn with a sigh of relief to plastic surgery to fetch the lineaments of their appearance closer to their heart's desire.

But Jenkins and Jenkins make a stronger point, one that needs to be taken seriously by immortalist utopians:

Reverse aging can't be a 'simple' matter of reversing DNA switches to express genes that have been switched off—nor of going back to a previous state of DNA, when it had more potential. In human beings, reverse aging would have to entail absorbing tissues that had already fully developed. Your joints would have to get smaller ... cross-linked collagen fibrils would break their bonds and become more flexible; scars and injuries would disappear and tissues return to pristine condition; fat would disperse from central storage collections to subcutaneous regions throughout the body; bones and muscles would regenerate and become dense; some glands would shrink and others would reappear.[22]

Still, even this apparently preposterous scenario—to which you can add new teeth budding in the gums and painfully pushing aside the last of your worn-down adult dentition, and presumably your bridgework as well—is not quite so farcical if the process is stretched out over months or years. And why shouldn't it be? Indeed, that's the way we grew from an egg to a mature adult; it took two or three decades. If full-scale rejuvenation takes a similar biological route, abetted by injections of gene-laden safe retroviruses, we may need to adopt a new cycle of extended life, with several years set aside to accommodate the discomforts and delights of growing younger.

Growing Organs in the Lab

We stand at the verge of regrowing whole new organs and replacing them without immunity-goaded rejection. You know something astonishing is happening when a major article on tissue engineering is the cover story of a business magazine, the 16 July 1998 *Business Week*, which lists dozens of biotechnology companies and university laboratories where such breathtaking work is proceeding apace.[23] Now is clearly the time to invest in surrogate organs.

In 1994, a 12-year-old youth, born missing the bone and connective tissue of the left half of his chest, had engineered cartilage implanted over his heart by surgeons at Children's Hospital in Boston (where Judah Folkman developed his anti-angiogenesis treatments). The new torso was held together by a matrix of polyglycolic acid polymer, a scaffold that biodegraded as the cartilage cells multiplied under the direction of added growth factors and specialised into their desired roles. The successful technique has been developed over more than a decade by paediatrician Joseph P. Vacanti and his anaesthesiologist brother Charles A. Vacanti, working with MIT chemical engineering professor Robert S. Langer. It is a stunning demonstration of the scale of repairs we will see in the next five to 15 years.

Skin suitable for transplant to burns patients has been available for some years, grown from snipped foreskins into flat sheets of any size desired. Now, though, three-dimensional organs are being constructed from degradable substrates seeded with cells and chemical promoters. Luckily, the environment of a living (if ailing) body knows enough to instruct the new cells how to switch on relevant genetic instructions and specialise to fill the role they find themselves in. 'Livers, pancreases, breasts, hearts, ears, and fingers are taking shape in the lab,' wrote journalist Catherine Arnst. 'Scientists are even trying to develop tissues that would act as drug-delivery vessels. Salivary glands could secrete anti-fungal proteins to fight infections in the throat, skin could release growth hormones, and organs could be genetically engineered to correct a patient's own genetic deficiencies.'

Bones are being grown in the Vacantis' bioreactors at the University of Massachusetts at Worcester, ready to be transplanted into workmen who lost their thumbs in industrial accidents. The researchers expect that the new thumbs will be working within three months after surgery. In mid-1998, a conference in Toronto set out to grow a whole human heart within a decade. Shaped knee cartilage will be grown in place, rather than replaced by synthetics. Urethral function, which weakens with age and leads to incontinence, will soon be augmented using bio-engineering. Reprogenesis Inc., a Massachusetts research firm, is in late-stage clinical trials, Catherine Arnst notes. Their process 'removes a few cartilage cells from behind a patient's ear, grows them in the lab, and then mixes them into a gel matrix. The cells are reinserted endos-copically where the urethra meets the bladder. There, they grow to bulk up the tubal walls.'

A bone-growth promoting substance, OP-1, stimulates stem cells placed in a calcium scaffold and grows fresh bone. It is even hoped that teeth and receded gums might be forced to regrow their enamel, since the relevant genes are already sequenced. Fractured spines, cause of paraplegia or worse forms of paralysis, might be mended, if current work with rats is transferable to humans. Many other growth factors are now known, and are under investigation, including human basic fibroblast growth factor (FGF-1), vascular endothelial growth factor (VEGF), trans-forming growth factors beta 2 and 3, purified growth hormone (somatotrophin), dissociable chromosome healing factor (CHF). Recent research shows that no growth factor maintains cellular growth if telo-meres have shortened too far. Growth resumes, however, in the presence of telomerase. Luckily, that does not imply unchecked renewed prolif-eration of tissues if such growth factors were introduced to an ageing body, or their dormant genetic expression kick-started by a drug. Organic development, after all, is a skein of preplanned events, governed by chemical gradients, the condition of surrounding tissues, nutrition, a hundred complex interactions. Once the growth plates at the ends of bones reach their ordained length, for example, they close. Reactivating telomerase and growth factors would not prompt acromegaly or any other pathological burst of bone enlargement.

In September 1998, scientists at the Wistar Institute in Philadelphia announced in *Proceedings of the National Academy of Sciences* a new mouse variety (MRL/MpJ) capable of 'rapid and complete wound closure that resembles regeneration'. Usually the tissue gaps caused by wounds are filled by proliferating a kind of generic scar tissue, but these mice also show 'recovery of normal architecture' *without* scarring. How so? The regenerative ability is apparently linked to several mouse genes found in the same place on a chromosome that in amphibians causes this effect. Wistar's director, Giovanni Rovera, was quick to stress that 'regeneration of an organ or a limb is controlled by many elements, of which these genes are only representing one'. Nevertheless, this is extremely encouraging news. Completion of the Human Genome Project could lead to identification of comparable chromosome locations in humans, which might be adjustable using the mouse genes as a template. Even if we have to do it piecemeal, it might be possible to stave off death until a more general cure for ageing is found—assuming that this goal is a plausible one.

An even more remarkable development was announced in November 1998, when two US laboratories reported that human embryonic stem cells, gathered from aborted foetuses and unused IVF embryos, have been cultured successfully. Papers in the journals *Science* and the *Proceedings of the National Academy of Sciences* describe how teams from the University of Wisconsin-Madison and Baltimore's Johns Hopkins University School of Medicine have finally learned how to grow these immortal cell lines in the lab. Because stem cells have the ability to mature into any specialised kind of tissue, this breakthrough will permit the construction of entire organs for transplantation. And because tissue evades rejection, a stock of these cells will provide ample donor sources—assuming the ethical dilemmas can be dealt with—for the ill, wounded and aged.

Very High-tech Solutions

An entirely different and even more optimistic scenario opens up when we take into account the spectacular new nanotechnologies expected by optimistic futurists such as Dr K. Eric Drexler and Dr Ralph Merkle of the Foresight Institute in the next 30 or 50 years. Nanotechnology (as I detail in my book *The Spike*) is the coming method of molecular manufacture and manipulation, using programmable probes and assemblers the size of viruses. For many of us alive today, that prospect is perhaps too remote, just out of reach. We must hope that incremental medical advances of a more orthodox kind can keep us alive and ticking long enough for molecular repair systems to come into play and perform the really *big* tasks— swarming throughout the body, nipping and tucking, correcting damaged DNA codons, fixing broken or cross-linked collagen strands, scavenging cholesterol deposits, resetting the expired instructions that leave the elderly with little energy, less hair and far too many wrinkles.

Another method for outwitting death and decay, still more radical, is the possibility of 'uploading' your brain structure into a supremely fast and commodious artificially intelligent computer, or constantly 'backing up' your memories into a plug-in chip that will survive with a kind of echo of you inside it even if your fleshly components keel over. That is not altogether ridiculous or abhorrent; we already have a great measure of redundancy in our brain, and can lose many neurons without any spiritual diminishment. Adding a kind of durable hard-drive as an extra module would surely enhance, not dehumanise, human possibilities.

If you happen to die before such a procedure is commercially available—and it's surely not to be expected, alas, until the middle or end of the twenty-first century—why, you can already buy a policy with a reputable cryonics firm and have your brain or whole body frozen, awaiting nanotech revival. That might be done either in a new body grown from instructions in your stored DNA, in your old body after a thorough molecule-by-molecule recondition (necessary to fix the awful damage that even careful rewarming wreaks on tissues), or uploaded into virtual realities inside a computer. Again, all these drastic possibilities are explored in detail in *The Spike*, so I won't cover the same ground here. In any event, it is important to understand that quite probably life extension and even physical immortality will emerge during the twenty-first century even if none of these other Promethean and still unproved technologies make it off the CAD-planning screen.

THE COST OF IMMORTALITY

Ultimately, then, we might expect to fix what is only contingently called 'ageing'—the damage and at last senescence that now accumulates with the passage of time—and find ways to outwit it. Certainly the human population explosion will become an even more pressing emergency, but we must solve that in any case. Perhaps the price of immortality, imposed by the community, will be elective or imposed sterilisation. Does this also imply a sterile cultural future? Of course not. Why should it? Remaining young for centuries or millennia is *not* the same as an eternity of sclerotic senility.

A youthful 300-year-old with a fresh body and undamaged brain will presumably be wiser, perhaps more set in his or her ways than an adolescent, but nonetheless without the crotchety traits the old exhibit due to their physical deterioration, and the frustrations and sourness that can entail. On the contrary—we will have time to learn and attempt all things—especially, as the optimistic novelist Robert Heinlein wrote, time enough for love.

Recall that editorial I quoted earlier. It ended, to my horrified disbelief: 'Dying (we think) is not all that bad.' Frankly, I can't think of anything to be said in its favour. I was no less taken aback to find similar opinions expressed in a 1997 *Scientific American* review by the fabled Leonard Hayflick, one of the world's premier biogerontologists and a pioneer in cell replication work. When an expert of Hayflick's standing speaks, one must listen with due respect—but in this case I disagree powerfully with his conventional opinion:

> *I have always worried about the enormous power that humans will have if we ever learn how either to tamper with the aging process or to extend our longevity—it is unclear whether people could cope with the psychological, economic, medical and cultural changes that would accompany vastly extended life spans, even should they prove physiologically possible … Although aging and death put an end to the lives of good citizens, they also make finite the lives of tyrants, murderers and a broad spectrum of other undesirables. Much of the continuing massive destruction of this planet and the consequent ills that this destruction produces for humans can be traced to overpopulation, a phenomenon that appears to show no sign of abating. Extending the life of a population that already strains global resources is, in the view of many, unconscionable. If the price to be paid for the beneficial results of aging and death is its universal applicability, we should all pay that price—as we always have.*[24]

This is a confused, sadly short-sighted estimate, in my view. True, society will convulse, in one way or another, as we learn to cope with extended and then indefinite life. That is no reason to prohibit the technology, or slow its development. Civilisation has already undergone shocking changes in mortality with the introduction of clean water and air, antiseptic childbirth, medicine and surgery, and plentiful nutritious food in the advanced sectors of the world. Would we be better off electing a kind of mad social amnesia, burning our medical books, closing all the hospitals, banning pharmaceuticals? Some say yes—well, let them retreat to their backward redoubts and live that way, but don't allow them to set society's health agendas.

What's more, Hayflick's premises are erroneous. Why should extending the life of a population, even one that already strains global resources, be 'unconscionable'? Buried in that critique is an indefensible generalisation: that longer life necessarily means more offspring. In rabbits or squid, perhaps that would be so. Humans have intelligence and foresight, when they care to use it. As Anders Sandberg, a Swedish transhumanist and neuroscientist, observes: 'This is another of the "classic arguments", and quite wrong (life span only changes the size of the population, not the rate of increase; if overpopulation is a problem then it has to be managed by family planning anyway).' We shall return to this issue of population dynamics at the end of this book.

Here's a more poignant objection: would it not be morally preferable to spend research money helping the sick and undernourished in the Third World, rather than expending a fortune extending the lives of the already-rich? Geneticist Michael Rose, at the University of California, Irvine, has nearly doubled *Drosophila melanogaster* (fruit fly) life spans by breeding a hardy variety with improved superoxide dismutase genes, and argues that we can do the same for humans. Still, Rose has remarked that while prolonging life is preferable to such audacious scientific feats as manned landings on the moon, it's 'not better than vaccinating all the children of the Third World'.[25]

This apparent conflict of interests, I believe, is rhetorical only. Must health efforts in one place necessarily be at the expense of those in another, when both could be funded? True, society's resources are limited (although immense), and need to be allocated with a due sense of proportion, but is there a more reasonable and satisfying way to spend part of them than in defeating death? Universal vaccination will save many millions of lives, and certainly should be funded—but each of those lives will only be saved for a maximum of a century or so. Life extension has the potential to save them, and us, for millennia.

What of Hayflick's cost-benefit analysis based on the prevalence and persistence of tyrants? Is it a strong argument for throttling the new science in its cradle? Surely not. If ever I've heard a cut-off-your-head-to-save-your-face pseudo-argument (and I'm not referring to cryonics here), this is it. Let us all die, to save a few from the heinous attacks of murderers? Kill everyone, by inattention, to spare us from the malevolence of dictators? In 1997, cryonics commentator Tim Freeman made the retort apposite:

> Suppose all reasonably free people got an aging prevention treatment tomorrow. So we'd eliminate the meaningless deaths of millions of talented and productive people in the US, Europe, places like that, but let's suppose for the purposes of argument that Saddam Hussein and the like would stay in power indefinitely, and the conditions of their subjects would be unchanged forever. Would that be a net win? In my opinion, it would be, even from a global utilitarian viewpoint instead of the obvious selfish viewpoint of a US resident.[26]

But is it even likely that the conditions of the oppressed could remain unchanged forever? Dictators and other bullies are always with us, for new monsters readily spring up to replace the old—who are often slain by the new, in any case. The remedy for death by terror and war is not involuntary 'natural' death for everyone, however peaceful. It is not universal, DNA-programmed mortality, but political awareness and action. Freeman added, astutely, 'Hayflick wasn't properly weighing the ordinary horror of aging that strikes the large populations that are not affected by tyrants. This is a common psychological error—people undervalue ordinary dangers, and overvalue unusual dangers.'

We're stuck, at the moment, with death's pain, loss and grief, and must make as decent a fist of it as we can. But in the longest term of the history of intelligent life in the universe, it will surely be the case— tragic, but blessedly brief in comparative duration—that the routine and inevitable death of conscious beings was a temporary error, quickly corrected.

three: life – Darwin's non-stop crap shoot

I do not see myself as a slave to the blind whims of evolution. Evolution is not a sentient process, and can therefore lay no claim to my obedience. I am a sentient being, and therefore by most ethical systems currently in use I should be free to take charge of my own destiny. That death may be a convenient means of speeding up the blind machinations of evolution means nothing to me. Frankly, it doesn't have any meaning regarding its 'rightfulness' in the past, either. Rights have to do with sentient beings. Is gravity 'right'? No, it just is. Therefore, when it gets in our way, there is no ethical reason for us not to strive against it.

Jeff Dee (1998)

Many people, to my amazement, denounce as immoral or repugnant the idea that we might enjoy indefinitely extended lives.

Where is the sweetness of life, they ask, without the stings, pangs and agonies of its loss? Life is the bright left hand of death's darkness. No yin without yang, and so forth. I have some sympathy for this suspicion (everyone knows, for example, that well-earned hunger makes the finest sauce to a meal), but not much. I'm not persuaded that simple dualistic contrasts and oppositions are the most useful way to analyse the world, let alone to form the basis for morality. Does freedom require the presence of a slave underclass? Are we only happy in our health because someone else—or we ourselves in the future—might die in agony from cancer? Let's hope not! I must state bluntly that this line of thinking smacks to me far too sordidly of a doctrine of cowed consolation, the kind of warrant muttered by prisoners with no prospect of eluding a cruel and unyielding captor, and with no taste for daring an escape bid.

It is a compliant slave's self-defeating question to ask *What would we do with our freedom?* The answer can only be *Whatever you wish.* Yes, freedom from imposed mortality will be wasted by some, life's rich spirit spilled into the sand, just as the gift of our current meagre span is wasted and spoiled by all too many in squabbles, fatuous diversions, bored routine, numbing habits and addictions of a dozen kinds. Others, bent by the torment of choice and liberty, will throw it away in terror, taking their own lives rather than face the echoing void of open endlessness. That would be their choice, one that must be respected (however much we might deplore it). For the rest of us, I think, there will be a slow dawning and awakening of expectations. People of exceptional gifts will snatch greedily and thankfully at the chance to grow, learn, suck life dry as never before. But so too, surely, will the ordinary rest of us.

With a span limited to a single century, a quarter devoted to learning the basics of being a human and another quarter, or even more, lost in failing health, it's little wonder that we constrict our horizons, close our eyes against the falling blows of time. Any one of us could learn in middle age to play the piano or violin, or master a new language, or study the astoundingly elegant mathematics we missed in school, but few manage the resolve. To make such efforts would be regarded by our friends as futile, derided as comic evidence of 'mid-life crisis'.

In a world of endless possibilities, however, where our mental and physical powers do not routinely deteriorate, opportunities to expand our skills and our engagement with other people, the natural world, history itself, will challenge all of us, including the most ordinary of citizens. Although it strains credulity right now, I believe one of the great diversions for many people in the endless future will be the unfolding tapestry of science itself (as well as the classic arts, and altogether new means of expression). By science I mean systematic knowledge tested against stringent criticism, and sought for its own soul-filling sake—plus knowledge as virtuoso technique, as the lever of power, enabling each of us to become immersed, if only as an informed spectator, in the enterprise of discovery.

SELFISH GENES, GENEROUS HUMANS

A quarter-century ago, Richard Dawkins, a winsome British zoologist with a gift for the sensational phrase, added to Darwin's century-and-a-half-old evolutionary scandal. The very title of his infamous first book, *The Selfish Gene*, was widely received as an outrage, a slap in the face to liberal sensibilities no less than to crusty conservatives who knew in their bones that this evolution chatter was all bosh and piffle. If ever the weasly term 'politically incorrect' had an apt use, that was it. Genes, selfish? But genes are the very elements of our heredity, aren't they, sacred sentences in the book of life? If they were selfish, if each DNA chunk were concerned for its own advancement and not ours, what nightmare universe did science propose as our habitat?

Dawkins was perfectly aware of the offence, and in his beautifully spoken, iron-willed way he repeated the affront at every opportunity. Like an archbishop of atheism, he was determined to preach his Darwinist gospel to the faithless and open their eyes, even at the risk of ridicule and persecution.

As it turned out, Dawkins was persecuted all the way to the bank. These days he holds a cushy post, the Charles Simonyi Chair of Public Understanding of Science at Oxford University. He is a master of memes, the swiftly replicating units of cultural life which he tells us infest our brains like computer viruses. The first words of *The Selfish Gene* seeded minds around the planet, enemies as well as friends. 'This book should be read almost as though it were science fiction,' Dawkins told us back in 1976. 'We are survival machines—robot vehicles blindly programmed to preserve the selfish molecules known as genes.' Many sensitive people still resist this portrait of reality. Ironically, Dawkins originally hoped to 'make everything about life fall into place, in the heart as well as in the brain'. His blind robots got to the heart, all right, and gave most people spiritual indigestion. For we are not robots, and it is the mark of totalitarians and barbarians to insist that we are.

SEEING REALITY THROUGH DARWINIAN EYES

In the revised edition, in 1989, Dawkins reported that 'the book's *reputation* for extremism has escalated', while its content 'has seemed ... more and more the common currency'.[1] Certainly the selfish-gene perspective has been well received by most experts (although some, like Stephen Jay Gould, disapprove of Dawkins' emphasis on the gene at the expense of the whole organism and the ecology it lives in). 'Rather than propose a new theory or unearth a new fact, often the most important contribution a scientist can make is to discover a new way of seeing old theories or facts,' Dawkins wrote. 'What we are talking about is not a flip to an equivalent point of view but, in extreme cases, a transfiguration.'[2] And that way of seeing the world captures in a lucid and uncompromising fashion the grand Darwinian vision of life.

For Dawkins was the very contrary of a barbarian. His public defence of science thrills with enthusiasm and wonderment. Might his PR triumph, his indelible memes, have done his case a horrible disservice? Should he and his fellow Darwinists step fastidiously back from such disturbing phrases, try to convince us that Darwinism sides with the angels after all? No. We must swallow the medicine if we are ever to be cured of muddle, bigotry, partisan superstitious cruelty.

As I stressed in the previous chapter, the single factor that persists down the river of time is genetic information. All the bright beauties of the day, all the sorrows of an eventful life, all the glories, triumphs, pains and tragedies of a human life are lost to the genome of the next child born. Yes, culture and society instil much of what makes us human (as indeed life with mother trains a kitten to be a cat), but finally what travels from deep past into deep future is a congeries of genes with a single common feature: they are *survivors*.

So 'flowers and elephants are, in effect, hosts to their "own" DNA,' Dawkins has remarked, 'in the same kind of way they are hosts to virus DNA.'[3] A virus infects us in order to multiply its own viral genome, using the machinery of our hijacked cells. To escape the host body and continue the race for more and more life, it tricks its host into sneezing, or vomiting, or bleeding, or pouring out life-threatening torrents of watery excrement loaded with infective copies of its genome. Luckily for host and virus alike, each species' genes work together in a mutually supportive package of instructions, which can be readily copied.

A virus, virtually a single strand of genetic material in a protein carry bag, replicates very directly. It just tells someone else's cellular machinery: *Make copies of me.* But the DNA in a leafy sea dragon or a tomato or a human has to take a huge detour, via sex: *Make me a body that can carry me to a matching complementary strand of DNA, and there make a copy of us both.* From the viewpoint of the genes (if they had one, which of course they don't), we are as ancillary to their purposes as our own discarded placentas are to us.

MERE RANDOMNESS?

Leaving aside the metaphysical voids this claim opens under us, does the explanation even make sense? Some critics continue to decry the alleged mechanistic crudities of such accounts, insisting that we are immeasurably more than the sum of our experiences. Oddly, evolutionists would agree in a way, for we are also the sum of our genetic heritage, the distilled wisdom of the ages. This will hardly satisfy those who yearn for an ineffable otherness, a profounder, limitless you.[4]

Regard the luminous natural world: could such majesty, such ingenuity, be the fruit of mere randomness? Of course not. Darwin's genius was to show that evolution works because its machine contains a ratchet, preventing an advance, once achieved, from slipping backwards and becoming lost.

If a gene helps its organism survive and breed, it will persist *because of that very fact* into the future. Even dud genes will survive for a time if they happen to share in a successful genome—but eventually chance recombinations will couple duds with duds, and their embodied carriers will perish, taking them down, one hapless creature at a time, into extinction. 'Darwinism is *not* a theory of random chance,' Dawkins reminds us, somewhat testily. 'It is a theory of random mutation plus *non-random* cumulative natural selection. Why, I wonder, is it so hard for even sophisticated scientists to grasp this simple point?'[5]

Consider the magical 'silken fetters' woven by spiders to catch their prey. This gossamer pours from spigots in the creatures' abdomens, and in some spider species must be eaten each night and hung out again the next morning. Webs are marvellously diverse. There are lacy spoked wheels, built with just the right balance of springy recoil and adhesion, and ladder webs stretching down from a branch or reaching up to one (independently evolved in, respectively, New Guinea and Colombia), and triangular sails, and bolos, and more. Could these miracles have 'just happened by chance'? Is this kind of laughable crudity, anti-Darwinist critics ask mockingly, to be taken seriously?

No, and yes (minus the editorialising). A computer program named NetSpinner simulates the evolution of web-making ability, starting with very simple strands, letting them catch or miss a bunch of simulated insects. When small mutations alter the plans of the next generation, patterns slightly better at seizing flies stay on as 'parents' in the next round. (In real life this selection effect would occur via enhanced survival chances of the web-spinners.) No hand guides the computerised evolution, yet within 50 generations lifelike webs stabilise out of rudimentary beginnings.

In other words, mindless computer simulations are now portraying the power of random chance coupled to ruthless, unguided selection for efficiency. Starting with a mathematical description of faintly sensitive skin tissue, a 1994 Swedish program by Daniel E. Nilsson and Susanne Pelger evolved (in two-dimensional schematic) the lens of an eye, that traditional marvel of divine handiwork. It took 363,992 tiny, plausible steps. So marine worms, molluscs and fish, with their short generations, could have 'invented' sight in less than half a million years. (Vision has been independently devised in nature, after all, by up to 60 separate pathways.)

Philosopher Daniel Dennett insists that mind and emotions, and all life's vast diversity, are the products of that single effect. A mechanical selection process, or algorithm, 'best accounts for the speed of the antelope, the wing of the eagle ... The underlying process always consists of nothing but a set of individually mindless steps succeeding each other without the help of any intelligent supervision.' Pushed to its limits, this is the troubling doctrine of the gene clawing past its fellows—although not so much a *selfish gene* as, more generally, a selfish *genome*. The ensemble of inherited genes that builds a single creature in a given environment does tend, after all, to hang together. But finally even this rule of thumb breaks down whenever it can. Mother and embryo, say, can battle to control levels of the growth-enhancer IGF-II, which exploits the female for the benefit of the foetus. There can be a genuine battle in the womb for control of resources, not the usual picture sentimentalists paint of Family Values.

GENETIC ESSENTIALISTS

The international Human Genome Project has been in full swing for a decade or more, mapping the sequence of DNA coding in both human and other model organisms. In humans, that's three billion letters of information (though much is apparently junk mail) comprising perhaps 100,000 different genes. I find it almost magically emblematic of the headlong rush of scientific prowess that James Watson, one of the two men who cracked the DNA code in 1953, should have been a director of the Project in its first four years. It's as if the man or woman who discovered fire went on in later life to oversee the invention of the steam engine. Ominously, Watson departed in 1992, apparently over the ethical question of who owns the human genome. He warned against a competitive, nationalistic approach to the new bioscience, where research groups try to patent strings of human DNA code.

Has Darwinism been hijacked by 'genetic essentialists' who sap our will to combat racism, feminist backlash, unchecked reproductive engineering and other horrors of the bad old days and the brave new world? That is the view of, for example, Dorothy Nelkin and M. Susan Lindee, sociologists of science partially funded, ironically enough, by the Human Genome Project. What is at stake for them isn't, on the whole, the real science, but intersections between lab and popular culture, so that 'the precise scientific legitimacy of any image ... is less important than the cultural use that is made of it'.[6]

DECONSTRUCTING LIFE'S TEXT

It is often said that the DNA inside our cells comprises a message, written (as the revealing terms put it) in a 'genetic code' or 'language of the genes'. Via a dizzying chemical virtuosity, DNA's four-letter alphabet and three-letter codon words construct our tissues, so each of us is 'written' into existence, within a specific, rich cultural environment, from that single recipe of 100,000 genes. Now we are about to read the entire document and, what's more, to edit what we find there and tack on a new ending.

The late Paul de Man, saint of American deconstructive literary and critical theory, once proposed that 'nothing, whether deed, word, thought, or text, ever happens in relation, positive or negative, to anything that proceeds, follows or exists elsewhere, but only as a random event'. Perhaps this idea is less breathtakingly mad in view of de Man's youthful pro-Nazi writings, which he never acknowledged but plainly wanted to wish out of existence retrospectively. But the idea that nothing happens except at random (a different point altogether from the indeterminacy of quantum theory, to which we shall return) also underpins a widespread theoretical claim in the post-modern humanities: that language is radically uncertain, so that no text can validly be given any authoritative reading.

If we are all DNA texts, might not molecular biologists and neuroscientists be advised to move their labs to the Literature Department? One specialist in the SV40 virus, Robert Pollack, actually does suggest that science has much to learn from current literary theory, and his argument is worth attending to (especially if, like me, you are a humanities theorist rather than a scientist). It's unlikely that Pollack would go as far as de Man in destabilising nature's text. While mutations and other forms of scrambled code are crucial to evolution, even more necessary is the strict specificity of each folded protein. Errors in copying or protein expression, as we have seen, are often lethal or at best hideously damaging.

Unlike human language, where our words are largely arbitrary (a sheep is a mouton is an oavtsa), 'a protein *is* the meaning of a DNA word', Pollack admits.[7] But since individual genes usually come in a handy range of options, or alleles, even close biological kin are an unpredictable mess of mix-and-match characteristics. What's more, the same gene cluster in a spleen cell will do a different job if a brain cell activates its message—an ambiguity that is, perhaps, not unlike context-sensitive literary interpretation. So science, no less than literature, needs to approach its DNA texts as documents with a *history*.[8]

THE EMPIRE OF THE GENES

So is history-conscious Darwinism okay, aside from an odious gene-centrism that has crept into the doctrine of certain powerful ultra-Darwinists? Genes might be potent, but surely not all-important. What about species, whose origin was supposedly explained by Darwin's great book? In its emphasis on the gene, ultra-Darwinism seems to ignore species, even dispute their existence. If selfish genes are the key players in this ancient contest, does this make the higher orders of taxonomy— the species, the families, the kingdoms of creatures—no more than optical illusions, like the observing 'self' that's composed (according to Dennett, as we'll see in the next chapter) of a multitude of mental gadgets? Palaeontologist Niles Eldredge, professor of biology at City University, New York, finds the DNA mystique less a political bother than a purely scientific one.[9]

For Eldredge and his sometime collaborator Stephen Jay Gould, standard Darwinism has taken a wrong turn. Not only is evolution's timetable made up of long stretches when nothing much happens, broken by catastrophes and bursts of frenzied change, but the current gene-centred emphasis is all wrong. It has lost sight of history. Genes evolve because the critters bearing them do well or badly in what we might call *economic* terms: according to how well their bodies make use of available energy and information. Competition for brute reproductive success, on this argument, is not the be-all and end-all of Darwinism, however fashionable that dominant notion has become.

No? Then why *are* we humans (to take the hardest possible case) the way we are? Is our history just a series of accidents? Arguably it's due, really, to sex. And sex is due to—what? The survival benefits of remixing the protein recipes? A recent suggestion is this: sex helps individuals fight off the parasites that infest all living creatures. Bugs, after all, breed very much faster than we do, so they adapt tremendously quickly to our individual defences. Only by churning up the DNA variety can a species hope to fight them to a standstill in a Red Queen's race where, as in Alice's Looking-Glass Land, one has to run as fast as one can to stay in the same place.

The upshot is a cascade of adaptive consequences, all the way up to the complex human brain/mind, which might have been evolved for its sexually appealing winsomeness, its social wit. If so, in zoologist and science writer Matt Ridley's piquant phrase, our 'virtuosity at everything from calculus to sculpture is perhaps just a side-effect of the ability to charm'.[10]

SEX AND GENDER AND THE WHOLE DAMN THING

So we arrive at gender (the variety of ways in which we experience our sexuality), as well as at sex (the basic reproductive structures). The great evolutionist John Maynard Smith has remarked wisely of troubling topics such as sociobiology, the theory that culture is at root biologically determined: 'I tend to find myself disagreeing most strongly with whichever side I talked to last'.[11] That surely can be said of those who investigate the politically hazardous topic of inherited cognitive and personality disparities between the human sexes. Gender plainly exists at the intersection of social pressures (including language) and our genetic makeup. Soon, in full-scale virtual reality, people will relate to each other convincingly in personae of whatever sex they choose (or indeed of animals or intelligent mists). Will optional presentation of self finally obliterate the socially constructed gulf between males and females?

Current evidence suggests powerfully that it will not. The language of the genes is inflected by hormones, especially those implicated in reproductive dimorphism—whether a child is to be a man or a woman, or some mix of the two. The evidence from subtle brain scans and long-term sociological studies argues that, statistically, some male and female brain structures tend to take alternative developmental pathways. Thereafter, in the summary of psychologist Camilla Benbow of Iowa State University: 'Women are drawn to people-oriented fields, men toward objects.' Too dangerous to think about? Too comforting to reactionaries? 'Women have been monstrously misused,' as Maynard Smith has noted, but 'the cause of women will not be helped by refusing to think dispassionately about the nature of sex differences'.[12]

An extreme example: famously, an identical twin boy was accidentally mutilated at seven months during circumcision. Snipped and tucked, he became a (rather tomboyish and pushy) girl. For many years her case confirmed the priority of culture over nature. Well, no. In 1993, it was announced that in adolescence she'd chosen to have plastic surgery 'to rebuild a scrotum and penis'. His brain, it's surmised, had been too 'masculinised' in the womb for her to live happily as a female adult.[13] (On the other hand, a 1998 study in the AMA journal *Pediatrics* of a similar case seems to provide evidence for the contrary, successful conclusion.)[14]

This new material—most of it collected meticulously by feminist women neuroscientists, not mad-dog patriarchs—calls for very careful study. Some women reject it out of hand as 'backlash' persiflage. Yet if these claims prove correct, even in broad tendency, the consequences will be sizeable. Horrifying though I find it, as an old '60s anarchist and feminist sympathiser, it might turn out that (in some respects at least) men and women are indeed complementary rather than equivalent. (One writer who has tried to draw this material together, avowedly even-handed and liberal, let his own guard down in a revealing slip. The female spotted hyena, Robert Pool remarks, 'sports a penis. Well, it's not *literally* a penis—it's merely a clitoris so enlarged that it looks like a penis'.[15] *Merely* ...)

True, there is a risk that the residues of historical happenstance and conflict, conditions that might have turned out otherwise and still could with a bit more effort, will be misunderstood as *Nature's Law* or some similar absurdity. You can find evolutionists deliberating on 'Polygamy and the Nature of Men' and 'Monogamy and the Nature of Women'.[16] This is exactly the kind of thing that discourse analysts such as Nelkin and Lindee repudiate: the gene as an 'essence of identity'. It might turn out that the historically distinctive behaviours of the sexes are indeed biased by their evolutionary histories, but such predilections are not instructions from above instructing us in our conduct. Careful analysis can help disarm fears of DNA totalitarianism. The scientifically informed philosopher Daniel Dennett, for example, is no less opposed to essentialism, and to what he dubs 'greedy' or wildly over-simplified reductionism, than any poststructural relativist.

BRAINS ARE FROM MA'S, EMOTION FROM PA'S

One place, though, where genetic explanation does now seem driven back into a form of essentialism, most unexpectedly, is the startling phenomenon of *genomic imprinting*. It turns out that the impact of parental sex on inherited character is not at all what old-fashioned bigots might like to imagine. So far the results have been demonstrated mainly in mice, but the genes involved are also found in humans. They imply that some traits are preferentially distributed unevenly among a couple's children. In brief, in the words of science journalist Gail Vines: 'A mother's genes play the dominant role in the development of the parts of her offspring's brains that are responsible for intelligence. The father's genes ... may shape ... the parts of their brains that influence emotional make-up'.[17]

This news must make immensely satisfying and amusing reading to those women who have suffered demeaning insinuations that they are surely responsible for any nervy, hysterical emotionality in their children, while stern fathers take credit for the keen manly intelligence that breaks through once the brats' moods and feelings are whipped into shape and brought under control. Quite the reverse is true, if this correlation holds up. It will not surprise those who have noticed how many clever children have smart, articulate mothers and quite ordinary or even lacklustre fathers.

Even before conception, genetic Darwinism is at work, labelling certain chunks of DNA—and not just in the sex chromosomes—to mark their origin in either mother's or father's germ cells, and to activate them only under particular circumstances. We need imprinted genes from both sex lines, since they do different jobs. That is one reason why radical hopes of creating an embryo by splicing together the haploid DNA of two spermatozoa (a baby with two genetic fathers) or two ova (a baby with two genetic mothers) won't work without *very* refined gene engineering, to the chagrin of some homosexual and lesbian couples. This, though, is not a futuristic gay utopia scenario—it was tried in 1984 using mice sperm and ova, and failed. In each case, a number of critical genes had been switched off in the DNA contributed by the two fathers or two mothers, so foetal development stalled. Mother-marked genes control early embryo growth, and the father-marked genes kick in with later development.

What happens if extra amounts of the genomically imprinted genes are added to the normal embryological mix, creating a chimera? Vines reports:

> *Embryos with an extra dose of maternal genes grew into fetal mice with big heads (and brains) perched on small bodies. By contrast, embryos with an extra dose of the father's genes grew into fetuses with huge bodies but tiny brains.*

Perhaps that sounds more like a coarse schoolyard cliché. Momma's boy nerds versus pinhead jocks. When the fine-grain detail was examined, the underlying effect was even more startling. Cells with father-marked genes drifted during development into the limbic system, the 'emotional' brain—notably, the hypothalamus and amygdala—but avoided the cortex and striatum, where fine movement is controlled. Mother-marked cells made the reverse choice, avoiding the evolutionarily more primitive regions of the brain in favour of the newer modules where planning is generated. Predictably, an entire book has already been published (by the perhaps appropriately named academic Christopher Badcock) arguing that paternal genes lay the groundwork for the Freudian *id*, repository of unconscious drives, while maternal genes do the same for the *ego*, or deliberative, conscious self. It is a notion as greedily reductionist and essentialist as anything in the ripest days of social Darwinism. For all that, it might contain some hint of truth.

But no more than a hint. The Darwinist view of life's panoply, from the earliest not-quite-living 'macros' to the finest works of art and science, is 'a breathtaking cascade of levels upon levels upon levels,' as Daniel Dennett observes, 'with new principles of explanation, new phenomena appearing at each level, forever revealing that the fond hope of explaining "everything" at some one lower level is misguided.'[18]

CRANES AND/OR SKYHOOKS

Not even the most optimistic reductionist has ever supposed, after all, that we interpret Shakespeare best by assiduously tracking the quantum states of each atom in each copy of *Hamlet*, or tallying the alphanumerics in *King Lear*.

Nor is human culture free-standing. A cornball but dexterous distinction of Dennett's is useful: using 'cranes' (evolved gadgets), minds and cultures are built from the ground up, not suspended in the empyrean from divine or mystical 'skyhooks'.[19] Humans, at a far end of those cascading levels of cranes, operate in an economy of memes (the term for units of culture borrowed, you'll recall, from Richard Dawkins) that rejig our purposes—though at the cost, perhaps, of ceaselessly battling genetic instructions embedded in our own cells and organs, and in those of the thousands of symbiotic bugs living inside us.

'In the 1990s,' as Nelkin and Lindee have noted, 'the popular rhetoric of DNA is as much about loss of control and acceptance of biological fate as it is about cure and the control of our future.'[20] So we need to step free of absurd popular iconographies and grasp the complexities of Darwin's truly hair-raising meme. It's important to acknowledge that most people who have adopted evolution as their working hypothesis of the cosmos, as I do, flinch from the horrors its pressures routinely produce. The only item in Darwinism's favour is that it's infinitely preferable to believing that some divine creator deliberately built the vicious ecologies we inhabit. Its strategies, as Dawkins has eloquently explained, are founded on the single fact that each of us is the end result of a long line of *ancestors*, beings who all successfully passed on their genes. Happily, not every product of random mutation and natural selection is vile. Such a filter can employ many strategies and tactics, including love, kindness, generosity, honour, artistic and scientific genius.

LOVE AND THE INNER INSECT

It need not do so, however, and a more searching gaze finds even at the heart of these humane values the same inhumane machinery that drives the praying mantis. The male mantis is eaten by the female as he copulates with her. Why doesn't he escape? Aren't his genes interested in preserving him? No, they're not, because they've found a better trade-off. As Randolph M. Nesse and George C. Williams have observed, 'he probably maximises his own reproductive success by donating his bodily protein to the female, who can use it to give more to their offspring'.[21]

This doleful tale is not a misogynistic fantasy from the divorce courts, just a horrible reminder that evolution conserves nothing but 'inclusive fitness': comparative success in making progeny by any means possible, including both 'altruism' toward other family members or gene-sharers *and* their heartless exploitation. If a strategy increases inclusive fitness more effectively by building minds and bodies exposed to disease, decay and death, and plagued by demons of unhappiness and strife, it will triumph over its sweet-natured rivals. Our origin in evolutionary contest is why we suffer agony and inflict it (as well as luxuriating in devotion and joy), and goes a long way to explaining the persistence of evil.

When such ideas were first systematised about a quarter-century ago as sociobiology, there was an outcry from humanists and other decent, fair-minded souls. Social Darwinism! Racism! Reductionism! Nazism! Those charges were usually wide of the mark. Indeed, most evolutionary forays into human affairs were done by left-inclined and compassionate academics. Today, as the human genome mapping project nears completion, there's an urgent need to reconcile neo-Darwinism with our tenderest cultural longings. E. O. Wilson, dean of sociobiology, now speaks of *consilience* and *biophilia*, terms with a sweeter, kindlier tang to them.

Nonetheless, for George C. Williams, professor emeritus of ecology and evolution at SUNY and one of the germinal figures in decoding the link between genes and behaviour, 'natural selection ... is an "evil" process, so great is the pain and death it thrives on, so deep is the selfishness it engenders'.[22] The title of a recent book by Williams—*Plan and Purpose in Nature*—is bleakly ironic, and his moral outrage at evolution's stupidity and cruelty untempered. 'Organisms,' he argues, with plenty of gruesome instances, 'show the expected stupid mistakes, the dysfunctional design features, that arise when understanding and planning are entirely absent.' Happily, as he urges, 'we can have some hope that our intelligent efforts to circumvent the evil can triumph over so unreasoning an enemy.'[23]

In short: recognising a process or proclivity as 'natural' or 'in the genes' does not endow it with righteousness. We are smarter than evolution—which is merely relentless—and we are more moral.

THE MORALITY OF DARWINISM

Some of those who once denounced sociobiology as racist or discriminatory now embrace it, for reasons just as dubious. Certain gay activists express satisfaction when lab reports suggest non-reproductive sexual preference might have a genetic component, although this in turn has been questioned.[24] It's hard to see, though, how that would or could matter. *Is* does not imply *ought*.

We now know, to take a horrible example, that 'the risk of fatal child abuse for children living with one non-genetic parent is 70 times higher',[25] even when contributing factors such as poverty, madness and drug abuse are factored out. This, awful as it is, turns out to be consistent with a common reproductive strategy in other animals, where males kill their deposed rivals' newborns (or cause their new mates to abort). Humans who act like beasts might, in fact, be acting like beasts. 'Animals,' Nesse and Williams declare, 'are inevitably designed to do whatever will increase the success of their genes, grotesque though the resulting behaviour may seem.'[26]

Of course, that particular strategy is not universal among humans; luckily, it's quite rare. How is that possible, if we are ruled by our inheritance? Because genetic regulation of behaviour is complex, and indirect. 'People are not controlled by some internal calculator that crudely motivates them to maximise their productive success. Instead, people form deep, lifelong emotional attachments and experience loves and hates that shape their lives.'[27] While the final arbiter of natural selection is an individual's fertile progeny—or close kin sharing the same genes—that brute fact cannot serve as a sufficient moral guide. Values, created by complex human brains, have burst free of the imperatives of the genes. Alas, our hard-wired animal inheritance has not. So we live in endless conflict between what is and what ought to be (whatever we decide *that* is, this week).

This fact is hardly news. The flesh is notoriously weaker than the spirit, and even the spirit has its own ways of going lethally haywire. Take a look at Bosnia or central Africa. Psychoanalysis and the arts are eager to tell us about mixed or hidden motives, most of them sexual one way or another. Evolutionary psychology and medicine, though, are starting to show us how to track down why and how these apparent aberrations are structured into life. These are hard doctrines to accept, however brilliantly apt their explanations. Their cynicism is extreme. Nothing sacred is left without its base account.

APE AND ESSENCE

At the extreme, we end up with a sort of vulgar gnosticism, embraced recently by a former celebrity PR man turned would-be guru, Howard Bloom: 'The nature scientists uncover has crafted our viler impulses into us; in fact, these impulses are a part of the process she uses to create. Lucifer is the dark side of cosmic fecundity, the cutting blade of the sculptor's knife. Nature does not abhor evil; she embraces it. She uses it to build.'[28] Oddly, even Bloom's defective version is less cynical than the mainstream varieties deployed by Dawkins, Dennett, Robert Wright, Nesse and Williams. These evolutionists stand by the individual, not the group, as nature's key.

It is the genes secreted within sperm and ova, and expressed in the magnificent complexity of bodies primed or biased to act in certain ways, that get selected. Which is why males and females are almost always born in equal numbers. How so? Regard the logic: if groups rather than genetically 'selfish' individuals were at the sharp end of selection, the sex ratio would be biased toward females, because the number of fertile females represents the limit on the number of children a group can produce in a given season. Actually, though, the contest of individuals ensures that whichever sex happens to be more numerous at a given time will have lower reproductive success. If much more than half the potential parents in the next generation are female, genealogies tilted toward boys will do slightly better, and so they'll thrive. Eventually, over time, this seesaw must reach equilibrium, and equality of male and female numbers will re-emerge.

Robert Wright has recently applied this kind of analysis to human values in an impenitent, even savage, demystification. It becomes evident that a full-bore deconstruction of traditional western values in the light of Darwinism is deeply disturbing. In a *coup de théâtre*, Wright examined Charles Darwin's own notably sweet and generous life for evidence of our shaped and shaping genes, and found it everywhere.[29] I reeled away from his unremitting cynicism and felt like scrubbing my brain out before I could eat off it again, but Wright is a moral and admirable fellow who searched for ways to reconcile selfish genes 'with the limitless compassion that is always, in fact, appropriate'.

THE EVOLUTION OF DEATH

Perhaps the most piercing questions we can direct to evolutionary storytellers are the ones we have explored already: *Why do we die? Why must we suffer illness?* Medical exploration has begun unfolding for us, as Randolph Nesse, George Williams and their colleagues have shown in detail, how sickness and health are usually 'compromises struck by natural selection as it inexorably shaped our bodies for the transmission of our genes'. This is a cynical interpretation of our woes, admittedly, but no cause for final despair. Even Dawkins reminds us that genes are, after all, not *entirely* selfish.

Here's the Darwinian crux again: every living creature is the end result of a relentless filtering process that has passed only *ancestors* whose genes built active bodies that thrived in their ecological environment. By the accretion of many small variations in phenotype, the agents at the sharp end of selection—the genes—are winnowed again and again by contest for survival in a world where many more individuals are born than can pass on their idiosyncratic genes (more precisely, alleles or gene-variants).

Such Darwinian filters operate wherever there is a contest among replicators for survival, but this fact can't tell us which strategies might work best in the arena. Clumping together cooperatively has its merits. So does defection. Game theory suggests that each of these strategies will come into play, perhaps with a measure of randomness to evade anticipation. Knowing this much can guide our understanding of mortality, but the details (and even the broadest generalisations) still require careful gathering of evidence and thorough analysis. Let us look once again at the strange phenomenon of menopause, and see what it looks like from the evolutionary perspective.

DIE YOUNG, AND LEAVE A BEAUTIFUL CHILD

We are finally getting to the point where all the threads we've examined loop into a satisfactory story, even if it is still sketchy. Indeed, several quite contrasting plots have been discerned recently, and it is not clear yet how they can be reconciled. It's worth trying, however.

Menopause—'the Change'—occurs, you'll recall, when pituitary hormones fail to activate repair and maintenance systems further down the track. It seems an absurdly early point in human life to shut down the reproductive system, since menopause begins thirty-odd years before many modern women are ready to die of old age. Surely the Darwinian dogma tells us that the sterile lose their place in the evolutionary queue. However healthy they happen to be, their success as *phenotypes* (or bodies) cannot win their genes a place in the next generation. They might even find fertile mates, but they have already lost the contest. There are well-known exceptions to this harsh economics, mostly among insects where sterile workers are selected because they share more than the usual proportion of genes with their fertile sisters, and aid the hive's prosperity. Humans don't have that advantage—except on a cultural level, where single celibates, dedicated homosexuals and other 'non-breeders' enrich the survival chances of their child-bearing fellows. Is that the clue we need to explain menopause's peculiarities?

Recall that the endocrine cascade of menopause starts when the last of the egg follicles in a woman's ovaries run out. This brute fact can be interpreted in several ways, some of them profoundly misleading. Without eggs, it is sometimes said, a woman is no longer at the sharp end of the natural selection process, so she can be discarded—left to die, literally. This, as we'll see, is exactly what happens to other mammals such as baboons and lions. A more accurate way of putting it is that, in practice, bodies built from genes which managed to keep them healthy and nimble until they were a thousand years old would have no reproductive value over bodies that drop dead of heart attacks at 50—

if they cease passing those genes on to descendants at 45. Evolution is a sieve for groups of genes, not for bodies as such.

One trouble with that stark account, which seems sound as far as it goes, is that it neglects the contribution of post- or non-reproductive family members who share the same genes. Think of two imaginary human tribes, each with several inter-marrying bands, who differ genetically in this respect. In one, mothers simply die at about 50 or 55, seven years after giving birth to their last child, conceived from one of their last eggs. The moment the youngest child in the family is reasonably self-sufficient, the mother (due to her characteristic blend of uncaring genes) gets cancer or heart disease and dies. In the other tribe, mothers remain reasonably healthy after an uncomfortable transition through menopause, and continue as productive members of their community, providing food and, as custodians of wisdom, hoarding and dispensing useful information that can make all the difference between life and death to many of their grandchildren and nieces and nephews for a couple more decades.

It seems intuitively obvious that the second evolutionary strategy will prove more successful. Indeed, it is embodied in a recent anthropological doctrine known as 'the Grandmother hypothesis', advanced in 1997 by Kristen Hawkes and James O'Connell, of the University of Utah, and Nicholas Blurton Jones, of UCLA, based on research among the 300 Hadza hunter-gatherers in hill country near Lake Eyasi in northern Tanzania. The healthy Hadza grandmothers foraged for food shared by their grandchildren, which permitted their daughters to bear and suckle new babies rather than expend energy and effort looking after their older children. It is true that acting as wise elder repositories of knowledge (memes, in short) would benefit the whole tribe indiscriminately. Hence, this memetic or cultural benefit might not have a direct and preferential effect on genetic reinforcement in the blood-lineage alone, and might not be selected for. Still, providing care and food for their daughters' children certainly *would* benefit their own gene line.

This is an endearing notion, and useful in offsetting the bombastic stress some male anthropologists have placed on the supposedly supreme virtue of hunting as a source of food (not that this has been taken terribly seriously for decades anyway). The problem with the grandmother theory, alas, is that it just doesn't seem to make mathematical sense. And in genetics, that is a very important hurdle. There's nothing abstract and arbitrary about the mathematical models used to test population genetics. The equations are clear and unforgiving, if used appropriately.

In September 1997, Natalie Angier reported in the *New York Times* that, 'Using mathematical models, Alan Rogers of the University of Utah estimated that a postmenopausal woman would have to double the number of children her children bore, and eliminate infant mortality among those grandchildren, to make menopause look like a sound strategy for propagating one's genes.'[30] That's an improbable improvement in life chances for any woman's descendants, even a hardy Hadza's. I suspect, therefore, that in this raw form, at least, the grandmother hypothesis has failed the test.

On the other hand, George C. Williams' model of inclusive fitness might yet save the day. I find it impossible to believe that the presence of at least some wise elders would make little or no difference to the survival prospects of a typical foraging tribe. Today, sadly, that's not really true. It is surely *nice* for small urban children to have the loving attention of their grannies, and convenient for hard-working parents to be able to turn over some of the task of child-minding. But none of this worthy contribution makes any difference, I think, to the number of children who survive to maturity and pass along the family's genes. Everything has got much more complex and confused with the coming of contraception, abortion choice, smaller families. If anything, it's those with the least resources and most disrupted family arrangements who are likely to have a larger brood.

Among other species, the evidence is ambiguous. A 1998 article in *Nature* by Craig Packer, a biologist at the University of Minnesota, St Paul, Marc Tatar of Brown University in Providence, Rhode Island, and Antony Collins of the Gombe Stream Research Centre in Kigoma, Tanzania, noted that

reproductive cessation has also been documented in non-human primates, rodents, whales, dogs, rabbits, elephants and domestic livestock. The human

> *menopause has been considered an evolutionary adaptation, assuming that elderly women avoid the increasing complications of continued childbirth to better nurture their current children and grandchildren. But an abrupt reproductive decline might be only a non-adaptive by-product of life-history patterns ... a systematic test of these alternatives us[ed] field data from two species in which grandmothers frequently engage in kin-directed behaviour. Both species show abrupt age-specific changes in reproductive performance that are characteristic of menopause. But elderly females do not suffer increased mortality costs of reproduction, nor do post-reproductive females enhance the fitness of grandchildren or older children. Instead, reproductive cessation appears to result from senescence.*[31]

The animals studied were baboons and lions. The distinction in life span is revealing. Female baboons who make it to old age become menopausal at 21, and survive an extra five years—a couple of years longer than needed for their last infant to outgrow dependency. Female lions live only 18 years, with menopause at 14, and their cubs need about a year of maternal attention.

Kristen Hawkes argues that this might not be as watertight an argument against the 'Grandmother hypothesis' as it seems, in evolutionary terms. Four-fifths of hunter-gatherer women, these days at least, survive beyond menopause, often into their seventies, so their contribution is marked. Lions, though, average only an extra three and a half years, and baboons only five.

TRIED AND TRUE

The basic argument here is that an organism that has stopped bearing young—or, in the elaborated version, can no longer contribute to the thriving of grandchildren—loses its relevance in the genetic struggle for survival. That is, we die after we have fallen too far behind in the struggle for genetic endurance. Accumulating errors exceed the ability of repair systems to proofread and fix them, even (to a lesser extent) in the sex cells. But this analysis, as we have seen, assumes part of what needs to be demonstrated—that older organisms automatically cease breeding reliably. While that is contingently true of complex creatures, as we have just seen, what we need to ask is: *Why should it be so?*

In other words, it is logically possible that an endlessly healthy, self-repairing and fertile couple could run a production line of their joint genes until the end of time, with an inclusive fitness function (in the jargon of the population geneticists) for their very own selfish genes which is superior to that of their rivals' gene pool. Both Darwin and Mendel would rub their hands with glee without a single censorious word of reproof. The only reason our earlier imaginary tribes came to the same grisly end was that the thousand-year-old women in the long-lived tribe were assumed, rather arbitrarily, to become infertile at about 45.

Why, after all, shouldn't the machinery of the body's maintenance systems keep working at peak efficiency forever? The accepted answer is that there are better strategic trade-offs for available energy and coding, trade-offs that allow the old to wear out and die. One reason this happens is that the world is a dangerous place. It's a *lethal* place, most of the time, in the wild—and the wild is where most humans have spent their time, until very recently indeed.

Another reason is that its deadliness keeps changing lanes, as viruses, bacteria and parasites frantically mutate. Sex happens to be a quite effective method for generating novel immune defence codes, blending two working systems into a fresh version, perhaps with some extra benefits. That makes sex a kind of genetic-immunological lucky dip, since many combinations are just as likely to make the creature they specify worse off—but evolution is a blind process, not a caring designer, and if one strategy is more successful than another, it is preserved, at whatever the cost to individuals.

THE HAZARDS OF BEING ALIVE

Meanwhile, the fittest person can be felled by a falling tree, be burned alive by a sudden bushfire, starve for want of food or drink in a lean season, catch some disease that even our sublimely devious immune system cannot counter. You can be seized by a feral animal, bitten by a spider or snake, fatally wounded in a territorial or sexually competitive fight with another human, or drown in rushing water. In other words, even if you remained as perfectly hardy as a 20-year-old, sooner or later you'd be cut down by accident or a newly mutated disease that evades your limited, preset immunity munitions.

If that were not the case, evolutionary sieves would not discriminate against older specimens. Why should they? If you can remain as alert as any adolescent and as wily as any mature adult, and retain sufficient fertility to produce a standard number of offspring at regular intervals, you are not at a Darwinian disadvantage compared with youngsters. They are just as likely to perish from all those awful enemies—perhaps more likely, being less experienced.

The balance tilts away from that perfect equality when subtle energy and coding considerations come into play. With every year that passes, there is an opportunity for cellular errors to occur and accumulate. That is inevitable, because chemistry is never exact. We turn over an immense number of cells, replacing old skin and other organs with new, and every time a cell divides to renew itself there's a chance for another blooper to enter your body. Sooner or later one of these faulty cells will prove to be critical, triggering a cascade of damage—notably, by switching off the $p53$ gene and allowing a formerly differentiated, law-abiding body cell to turn into a wildly proliferating outlaw cancer cell.

The body is forgiving where it can be, designed to work around minor failures. As time passes, we build up a legacy of slight damage, like worn bearings ready to cause an accident if too many things go wrong in a machine at once. What's more, some mutations will occur (at random) that create an unplanned early *advantage* to a growing individual, whether as a foetus or a child, but with its own *delayed cost* built in. Suppose a slightly elevated level of a given hormone gave a baby a better chance of fighting the infections of infancy, or processing available nutrients, but the price was an increased cancer hazard after about age 50. People passing along that allele, or gene variant, would have babies who thrived a little more than those of their neighbours. A new generation would contain more of these people than you'd expect, all things being equal. Then their parents would die of cancer, perhaps to the astonishment of the rest of the long-lived community—but their mutant genes would live on in the young, and continue to creep into the community through inter-marriage.

The tragic upshot is this: since many subsystems of a living physiology can vary in their effectiveness, sheer statistics will ensure the growing prevalence of more and more alleles that contribute to childhood thriving but hurt or even kill off the parents. This hangs on the assumption that some metabolic mechanisms especially beneficial to the infant, say, will turn and bite later on. Perhaps a chubby cholesterol-rich kid initially does better than the lean and hungry look, but arterial accumulation slays them after 50 fertile years. What's more, as George C. Williams has noted:

> There need not be any actual early benefits for senescence to evolve. As long as selection is more effective at suppressing unfavorable effects early in life than later on, adaptations early in life will be more effectively maintained than those at later ages. So that eternal-youth population, if it ever existed, would be unstable. Natural selection would quickly enhance youthful fitness at the expense of old age, and senescence would soon evolve.[32]

THE HAZARDS OF BEING CONCEIVED

But is exposure to free radicals and cosmic rays for 40 years so much more damaging, even lethal, than for 20 years? My hunch is that the machinery that keeps gonadal DNA clean and shiny is specialised and too expensive to share around generally. William D. Clark, Emeritus Professor of Immunology at UCLA, notes that germ cell DNA, peeling open during reduction division for copying, 'is readily accessible and easy to repair, and germ cells are literally loaded with the equipment to do it: DNA-repair enzymes.'[33] Even these heroic measures do not always work even for the purposes of protecting against explicitly reproductive damage, of course.

Down syndrome mental retardation, for example, is an embryological glitch that impacts on about one pregnancy out of 1000. It is usually caused by a surplus chromosome 21—Down embryo cells have 47 instead of 46 chromosomes—mostly acquired while eggs and sperm are being constructed prior to conception. In a few cases, however, damaged eggs have been waiting like a timebomb from their earliest formation when the mother-to-be was herself still a foetus.

The details are complicated, as always, since there are at least three ways foetal development can be diverted into the Down pathway. The most common (95 per cent) is trisomy 21, where a complete extra chromosome 21 from either mother or father stays stuck to its original partner during what's called meiotic reduction—the splitting of an adult cell into two, each sharing half the normal DNA complement. Meiosis allows the contributed half-shares of both parents to combine into a complete genome, but sometimes reduction division goes wrong and too many or too few chromosomes pass into the queued germ cell. Then the wrong mix of proteins is made by the developing foetal body—an excess, in the case of Down syndrome—and deformity of mind and body are the usual result. Trisomy 21 error mostly occurs during the ripening of eggs rather than in sperm; up to 97 per cent of Down syndrome errors derive from the maternal line.

As is now widely known, parental age is critical to the likelihood of producing a Down baby. That is true even of the fathers. A French study by Patel in 1995 showed that of pregnancies fathered by men aged 40 or older, about four in 1000 were affected by Down syndrome—three times higher than the rate for fathers under 35. For women, the chances are even more marked. Incidence graphs out a J-shaped curve, with few Down babies born to mothers 19 or younger (one in 1850 births). Then the curve starts to rise, to chances of one in 350 at age 35, and a sharp upturn to one in 20 for mothers over 45 years of age—due, presumably, to the age-impaired germ cells of older mothers, and a few older fathers, falling into meiotic error.

This sort of issue, once totally irrelevant except to biologists and the parents involved, is suddenly rather more pressing, as we move into a world where young women can act as womb-surrogates for embryos from older parents, and where quite elderly women can carry children if they are first given corrective hormone treatments to overcome menopausal shutdown. Fortunately, fairly reliable tests for trisomy 21 can now be carried out early in pregnancy, allowing for termination if that course is chosen.

At least in the short term, until the genome is completely rewritten to ensure immortality, enhanced longevity may well require the physical introduction of cleaners and sweepers at the nano scale, capable of inter-dicting such cellular and sub-cellular copying errors. They will get their purity guarantees, and perhaps their virus-scale power supply, not from fallible DNA but from outside the body: from culture's high-tech memory and apparatus, rather than directly from genes. Eventually, we might extend our bodies' ability to repair themselves, keep themselves young, by devising artificial chromosomes or plasmids (loops of DNA) distributed to every cell in the body via some kind of tailored and benign retrovirus. This could even be spread by a new breed of 'biological anti-terrorists', so to speak, as a kind of worldwide plague of immortality.

HOW LONG COULD IMMORTALS LIVE?

It sounds like a contradictory question but, as we've seen, death by accident and predation will scythe the ranks even of the otherwise deathless. If all in-built known causes of mortality were eliminated—genetic disabilities, common diseases, wear and tear—people would still die by misadventure. In 1988, the cryonicist Hugh Hixon published a fascinating analysis of the prospects for an otherwise immortal popula-tion. His bottom line, offered with all proper reservation, was astound-ingly optimistic.

Let's look first at what the biogerontology specialists have said. Leonard Hayflick, in his 1996 edition of *How and Why We Age*, took a conservative posture. Eliminate all present causes of death, he said, and we would get 'the hypothetical ultimate curve', a Gompertz slump tugged upward until it was almost rectangular. 'The world would then be one in which no one died from causes now written on death certificates ... no one would die young; all would slip peacefully away during a narrow span of years, say between age 110 and age 115.'[34] It was a view put earlier, in 1981, by James Fries, and vigorously debunked by Professor Roger Gosden in 1996. 'Most evidence is to the contrary,' Gosden stated, 'and it is nonsense to separate the underlying aging process from its manifestations. If anything, we are currently gaining more years of disability than of vigor.'[35]

That bleak opinion is undoubtedly correct, but we might well be in a transitional epoch. Simply abolishing the current causes of death will just be the start. We might expect a great deal more once medications are available that draw upon antioxidant and telomerase research, for example. Allow the possibility, then, that more is achieved than the removal of all known diseases and disorders. Suppose we have an enhanced immune system and other repair gadgetry—either genomic or nanotechnological—that keep us as young as we wish. What then?

In 1964, the British gerontologist Alex Comfort (in his study *The Process of Aging*) estimated an average cut-off date of 600 or 700 years, although he did not indicate how he arrived at this interval. 'If we could stay as vigorous as we are at 12, it would take about 700 years for one-half of us to die, and another 700 years for the survivors to be reduced by one-half again.' Hixon disputes this figure as pessimistic, based on a death rate from a period early in the century when childhood mortality figures were worse than today's. His own analysis, based on standard mathematical methods, uses the same idea of 'half-life'—the time during which half a population is still alive (or in the case of radioactive particles, where the measure is also familiar, during which half the atoms have decayed into stability.)

Using 1981 data from *Vital Statistics of the United States*, Hixon removes all but death by accident and homicide (assuming that suicide is a preventable mental condition), and finds '41.9 deaths per 100,000 in the white population (64.9 for males, 19.5 for females). Which gives us a *half-life* for our population of 1654 years.' In other words, if you are one of these immortals, you'd have an even chance of surviving more than a millennium and a half!

That estimate, however, is *still* very conservative, since it assumes that 1981's fatality statistics are valid for all time. As Hixon notes, we already have the technology for medical pagers linked to the Global Positioning System, able (in principle) to fetch emergency aid to accident scenes anywhere in the world at maximum speed. In a long-lived world, you can be assured that money would not be stinted in servicing this facility. The main risk of permanent death in an 'immortal' world would be massive cranial damage, the extinguishment of the neural basis of consciousness. Hixon guesses that this might be just one case in a hundred thousand. If so, you'd have 'a population half-life of *69,315 years*.

However,' he adds instantly, and wisely, 'anyone who quotes this figure without including a statement of its very speculative nature is on their own.'

This is a dizzying vista, but still very far from the prospects of foreseeable technology. As we shall see in the next chapter, mind is just the operation of the brain, and there is no fundamental reason why the brain should not be copied neuron by neuron—backed up, as it were—and its mind reawakened inside a computer, or merely expanded by plug-in chips. The posthuman world that some scientists predict for the latter part of the twenty-first century is almost beyond current conception and evaluation, but we can be sure that life in that epoch of exponential change will be like nothing seen up to now. These are the Promethean prospects I explore in *The Spike*, and they need to be taken fully into account in any realistic assessment of a humanity freed from biological mortality. If life escapes from its organic shackles—and I am speaking here of a purely scientific, engineering prospect, not a religious or New Age one—then we might genuinely find ourselves or our children living infinitely strange, challenging lives, not doomed to inevitable death after a century at best of increasingly frail life.

AGAINST NATURE OR BEYOND NATURE?
The question of 'interfering with Nature' does not oppress me as much as it seems to exercise many others. Perhaps in part that is because my mother didn't just have one genetic brother, the only uncle I knew as a child, but several other half-siblings. Yet her father had only one wife, and to the best of my knowledge he never had children by another woman. Even so, my mother's cousins were closer to her than usual by the rules of blood kinship. The tangled web had a curious source: my grandfather was a clone.

In a way I have been aware of this since childhood, where I first met his identical twin brother. Grandfather Chris lived in Melbourne, a couple of suburbs south of our small working-class home. Dan, the more prosperous twin (he'd been a cavalry Captain in the First World War, while brother Chris had been a gunnery Sergeant), lived in Sydney, and I do not think I had ever heard of him until the creepy day I came home from primary school and greeted a familiar, beloved man *who didn't know me* and denied that he was the grandfather I clearly saw him to be. Laughter all round from the adults—but I was distressed by the uncanny duplication.

Maybe it's that sort of unease which makes most people nervous, even revolted, by the prospect of human cloning and other disturbing examples of what Lee Silver, a Princeton professor in the department of molecular biology, ecology and evolutionary biology, dubs 'reprogenetics'. If identity itself is up for grabs, what is there to hold onto in an uncertain world? And if the sanctity of individual human life can be slighted at the very source, what moral or ethical underpinnings can we cling to?

Such concerns, I think, fuel the persistent anger and alarm in the debate over abortion. The debate has grown complex and ramified, but tends to reduce to an emotional stand-off. Some hold that human personhood begins at the moment a sperm enters an ovum and the parental DNA fuses into a unique living creature. Others argue that an embryo, or even a foetus, is not yet sufficiently complex and independent to claim the rights due to a human person. Which attitude you adopt will govern your reaction to today's drastically new biological developments.

Having created the enigmas of genetic profiling, bio-engineering, and now cloning (very soon, surely, of humans as well as sheep, cows and other mammals), can rational science help us unpick these ethical issues? Yes and no. Our received values are themselves often based on the best available scientific opinion—of the fourth century BC, or the thirteenth AD. Is abortion murder? Following Aristotle, Thomas Aquinas taught that the male embryo became a human being only at six weeks, while the female had to wait until 13 weeks. These were the dates when God infused a spiritual essence or soul into the raw matter of the developing flesh. More recently, Christian theologians have updated their science—to about the middle of this century. Human life, it's usually held, starts at conception. This, though, cannot be correct. In fact, the embryo does not combine the parental DNA contributions until the two-cell stage, on the second day. Even after that, each cell in the replicating ball of tissue is exactly identical until the fourth cycle of doubling. Then the 16 cells split into an inner bunch that will specialise into a baby and an outer shell that grows into a disposable placenta.

In various ways, the developing tissues with their unique DNA pattern can go astray. Sometimes they will split and restart their developmental program, yielding twins or triplets or quads—which are literally clones. The children of Chris and Dan share *exactly* the same paternal genetic inheritance. We always thought of them as cousins, but they were really brothers and sisters with different mothers, absurd as that sounds. Twenty years ago, Sherman Silber transplanted a testicle from Terry Twomey (a name that sounds suspiciously like a humorous pseudonym) into his sterile twin, Tim. The child of Tim and his wife Jannie is genetically indistinguishable from any that Tim might have fathered if he hadn't been afflicted with a rare developmental disorder that destroyed his own testes. So was this a kind of weird high-tech incest, or *in vivo* cloning? Presumably not. Yet Tim's son is also certainly the brother of any kids Terry has under his own steam with his remaining testicle.

How Many Parents, How Many Souls?

Modern reprogenetics multiplies such conundrums dizzyingly. You think you know what a father and mother are? Whose child is whose? Suppose you took ten very early embryos from as many random couples, separated the eight cells of each, and encouraged each of them to grow into a healthy baby. This would require most of them to be implanted into surrogate mothers, of course, and not all would survive. Still, in principle we could end up with ten sets of eight cloned babies—octuplets.

But go back a step. With the 80 cells segregated neatly, pick one cell from each of the ten Petri dishes and allow them to fuse together into ten entirely new embryos, called *mosaics*. This is not fantasy—using mice and other mammals, it's common procedure in the labs. Now you might have ten babies born, each of them with numerous parents. Here's the puzzle for theologians: if each original embryo has its divinely infused soul, how many do we have when they are split into 80 cells? Indeed, how are the souls reassigned when the cells are randomly recombined?

This is not a silly or blasphemous thought-experiment—a hundred natural errors a year cause twins to merge into mosaic individuals who usually grow up without subsequent problems. Do they have a double helping of souls? How does this impact on your moral evaluation of abortion?

The twenty-first century's molecular revolution will pose many wilder possibilities, such as children created for gay couples which share the heritage of both (although, as we've seen, genomic imprinting will need to be tweaked). Genetic profiling will allow parents to know broadly in advance how any given embryo would likely develop (mapping the predictable parts of the gene-programmed course on a computer display, portraying him or her at various ages), so that they might launch the life of one embryo instead of another. Eugenics? Perhaps, but not as it has been stigmatised by association with Nazism and other abominations.

The outcome, however, might be a growing genetic apartheid of rich and poor—what Lee Silver has dubbed the enhanced GenRich and the hapless Naturals. Some will embrace this fate as the fruit of the dream of consumer capitalism. Others will feel a chill. I don't see it happening, actually, because other technologies such as artificial intelligence and nanotechnology—perhaps necessary anyway to foster advanced repro-genetics—will make these benefits available to everyone, at least in the developed world. But it does seem highly likely that the human species is on the verge of branching into something old, something new, something borrowed and (as critics will insist) something bleugh.

'Meat Machines' and Other Dangerous Metaphors

We have ventured far in this survey of the ferociously contestatory world of genes and their large-scale manifestation, bodies and minds. It might seem that such talk inevitably degrades everything of human worth, both the beautiful and the terrible, reducing all to a crude mathematics of battle. It can even be claimed (and has been, all too often) that a Darwinian analysis of life asserts insanely that we do not really love our children but only go through robotic motions because they carry half the genes we do. This is a dreary parody of subtle evolutionary thinking, and needs to be set straight wherever it raises its duplicitous head.

If we are in some measure tools for our mindless genes, then we are tools with passion, reason and a powerful sense of free will. This last might seem paradoxical in beings supposedly governed by the rules of a deterministic universe, but as we shall see in a later chapter it is not the paradox it seems—or if it is, that is only because the cosmos is both very much greater (composed, arguably, of a multitude of universes that overlap and interfere with each other) and very much smaller (built from quantum particles and fields that can feel each other's presence instantaneously across the whole visible breadth of the cosmos). There is certainly room enough in such a glorious universe, to repeat Robert Heinlein's hopeful phrase, for love.

At the level of complexity represented by human 'meat machines' (a disturbing phrase—but in fact we *are* made, for the moment, of the same materials as meat) the tools for survival and prosperity include love and hatred, tenderness and rage, a hunger for companionable fun and a passion for quiet contemplation. Of course we genuinely like our kids, on the many occasions when we're not driven up the wall by them. We will give up comfort, sleep, even life itself if necessary, to spare them pain and danger. Science helps us understand how this is possible in a Darwinian universe, which is the kind we inhabit.

To work in bodies, genes need to be 'good companions', jostling in coalitions where the whole really does provide better pay-offs than the parts separately. Besides: chosen human values really can lift free of the dictates of the genes. But that opens a whole new dimension to the debate, signified by Dawkins' word 'memes', thought-viruses that infest human brains (although that is too loaded a term, since many of these 'infestations' *comprise* the structure of our thoughts and beliefs). Darwinism has now shifted up a notch; we genetic individuals, from one perspective, are the cellular substrate of the ecology of memes. If this claim seems preposterous, look again at Bosnia, or at the cumulative glories of the history of painting.

Memetic conflict of an especially crude kind very nearly destroyed the world in the second half of the twentieth century. It's possible that we were saved from nuclear holocaust when Carl Sagan and his colleagues spearheaded an already forgotten but crucial scientific PR campaign for the 'nuclear winter' hypothesis—the terrifying discovery that nuclear war would probably ignite the world's cities and forests, sending aloft vast plumes of soot to block the sun's light, plunging us into a catastrophic ice age. Since then we have found fresh horrors to alarm us: ozone holes, greenhouse effect, greedy destruction of the world's forests. Even as he was dying of cancer, Sagan and his literary collaborator (and wife) Anne Druyan tried to explain us to ourselves, to find—in our Darwinian links with the rest of the animal kingdom—the key to how we got into this mess and how to get out.[36] The risk, inevitably, was anthropomorphism, treating animals as if they have human purpose and foresight. Admittedly, there's evidence that bonobo chimps, at least, do possess a measure of selfhood. And we tend to deny these links with all our might, perhaps because we have often acted so disgustingly toward our primate kin.

BESTIAL SCIENTISTS

In 1986, for example, when animal rights activists broke into a US AIDS-research lab they found infected monkeys and chimps caged in literally inhuman (if clean) conditions. Roger Fouts and Jane Goodall, famous chimpanzee specialists, visited the nightmarish site. 'They had these chimps in metal boxes. One wasn't even rocking any more,' Fouts reported to Pulitzer-winner Deborah Blum. 'It was like those children you see in Somalia, that blank look ... And the vet said, "See, she's not screaming", and he told the tech to take her out. "See, she's just fine." They were holding her like she was a typewriter and she was just lying there.' The official observed cheerfully 'that they must be reassured by what they had seen'.[37] Goodall simply wept. Of course, one weeps as painfully when one's beloved pet dies, yet nobody believes dogs or cats have language-level consciousness. But I wept, too, reading that tale of ordinary atrocity, and of other sickening mutilations of animal minds and bodies (however well intended their medical purposes). Such reports press upon our complacency and human chauvinism. They bring our blood to the boil at the cruelties of—it has to be admitted in such cases—stupidly heartless science.

Scientists, like the rest of us, really are bastards sometimes, and the self-interest shaping their professional disciplines shields them from knowing it. William Jordan has written of a group of young biologists, himself one of them, through their training in the early 1960s, smashing rats to death (neatly), sawing open living turtles (painlessly—maybe), stabbing thermometers into mice 'nicked and bloodied' by human hair clippers. 'We have become calloused,' Jordan confesses. From day one, Jordan says, students must be made aware that 'the goals of science ... carry a price tag. The price is humanity. Each sacrifice, each dissection, each surgical incision is drawn on the account of good and decency ... Then teach some small gesture, some little expression of thanks to the creature ... Say a small prayer for the souls of us all.'[38] Even if, like me, you don't give any credence to the soul as an independent spiritual organ, you know perfectly well what he means.

One thing is clear: despite our apparent governance by language and culture, we share most of our genetic code with the chimps. That DNA-coding constrains and enables our feelings, our behaviour, our very sense of self and other. Science now shows us why such propensities 'are almost reflexive, why they should be so easy to evoke', as Druyan and Sagan have observed. 'But we cannot wait for natural selection to mitigate these ancient primate algorithms.'[39] As a political message, that is a far cry from noxious social Darwinism. If it is even partly correct, we are well advised to gaze into the shadows of our ancestry for hints on the emerging shape of our posthuman future, for we shall surely carry much of that baggage with us in our emulated and uploaded hearts.

Is Racism in our Genes?

How did our long Darwinian ancestry impact on those common features of all societies that get lumped together, along with various local biases, in the phrase and concept 'human nature'? Certainly evolution has shaped us all, wherever our ancestors first settled in their diaspora from the African plains, but it has been shaped in turn by geographical factors we often overlook. Consider the dread impact of white settlement on indigenous Australians. It is marked by a strange, terrible and instructive asymmetry.

Some 200 years ago, a handful of whites arrived in Australia, bearing in their tissues a starter-kit of disease pathogens that roared like a lethal fire through the 300,000 black people most of whose ancestors had lived in near-isolation on the island continent for upward of 50,000 years. Yet when a few Australian Aborigines visited Europe by return boat, no equivalent epidemics gutted London or Paris. Nor did the convict settlers perish from unknown local microbes. Why not? Something eerily, horribly similar had occurred 300 years earlier, when Europeans entered Mesoamerica. The indigenes fell like wheat, slain not so much by the steel and gunpowder of the invaders as by the germs they carried, all unknowing. Yet the Spaniards did not die in turn from Aztec or Inca bugs. What can possibly explain this fatal imbalance?

Europeans gained their immunity to a range of cruel pathogens from tens of millennia of drawn-out slaughter. Generation after generation were culled in a monstrous but almost invisible Darwinian contest of infection and adaptation. Waves of pathogens killed and killed, and the survivors limped on, passed down genes that gave the next generation a small edge in the long fight. So by the time the Spaniards arrived in the New World, their ancestors had paid in hundreds of centuries of pain and death for their invulnerability. The luckless native Americans, isolated for more than 10,000 years, copped the whole process in one appalling hit. Perhaps no more of them died than had perished in Europe from the same diseases—but all were felled in an instant. That compression of time's miseries obliterated almost the entirety of their history. In the heartbreaking extreme, the alien diseases 'spread from tribe to tribe far in advance of the Europeans themselves', as physiologist Jared Diamond has noted, 'killing an estimated 95 percent of the pre-Columbian Native American population'.[40]

I don't know if such accountancy is the best way to portray a vast human tragedy, but it provides a kind of concentrated image of 'geographical possibilism', as Diamond's critic Timothy Taylor has dubbed it, which Diamond himself calls 'the fates of human societies'. Why have white cultures developed machine technology, medicine, abundant consumables, while 'non-European' cultures failed to do so? The typical answer, usually unspoken, is racist. We see its continuing power in today's politics. Blacks and other allegedly 'primitive' people, according to this easy prejudice, are stupid, lazy, shiftless and probably driven by the beast within. Diamond's own decades of friendship with New Guinea highlanders convinced him that if anything these folks are smarter and quicker than Europeans.

Why, then, was it whites whose history ransacked the world and built a powerful, global scientific civilisation? The usual answer is that they owned advanced weapons and industry, and hosted lethal germs. They possessed horses, plentiful food, and writing. But how did whites from backward Europe chance to be the first kids on the block, considering that New Guinea farmers were building irrigation systems 9000 years ago? When humans had thrived in Africa for many hundreds of thousands of years before the great diasporas into Asia and Europe?

Early cultures gained the power to overwhelm and replace others when they discovered the arts of planned annual food production (rather than opportunistic gathering). Grow your own food, in herds or by farming crops, and you access reliable surpluses. Populations boom. Specialists can devote time to the arts of war and peace. Bands give way to tribes and chiefdoms, conflicts can no longer be adjudicated within the extended family, politics emerges, finally writing and the extended memory systems that permit complicated trading over vast distances, the rise of new technologies ... All these factors hang from each other, and modulate local cultures in a thousand ways. Still, we can't avoid asking what it is in the histories of different cultures that allows one of them— the northern Sino-Tibetans—to take command of China, while restricting another to a single island (like the placid Moriori, massacred by 900 Maori boat-borne warriors in 1835).

The answer might be found in the disposition of the continents and islands on our tilted, seasonal planet. Here is something we urban dwellers forget all too readily: despite the global village, the world is not all alike. Fresh, delightfully various foods at the supermarket must be fetched from places very far away. The world is striped and zoned. Animals and crops from *here* will not thrive *there*. The large creatures we domesticate to pull our carts, or to eat, and those other tiny invisible animals that eat and breed in us, are not native to the whole world but prefer certain localities. That is the basic Darwinian perspective.

LOOKING SIDEWAYS AT THE EARTH

Picture the horizontally striped globe, and something jumps off the map: the orientation of the great land masses. Europe and Asia (Eurasia, in effect) spread endlessly east-west. If crops and beasts of burden evolve at a certain latitude in the Fertile Crescent, they will thrive in a vast ribbon across Eurasia. The Americas and Africa, by contrast, run along a polar axis. Mediterranean climates are separated by tropics and deserts, forests by formidable mountains. Today we can transport livestock across those barriers, but in the epoch of first human cultures they were effectively impassable. Llamas in the Andes could not be harnessed to the wheel invented nearby in prehistoric Mexico. Worse, late-arriving humans in the Americas and Australia had met walking larders, fearless megafauna they hunted into extinction—thus killing off the big animals their descendants might have used to pull ploughs and chariots.

Once population density and growth began their runaway in Eurasia, fuelled by domestication of animals for food and traction, the subtle horror of new disease was launched. For the big infectious killers evolved from diseases specific to animals—flu (pigs and ducks), measles, small-pox and TB (cows), malaria (birds). Farmers are terrible foes, as Diamond remarks, because they 'tend to breathe out nastier germs, to own better weapons and armour, to own more powerful technology in general, and to live under centralised governments with literate elites better able to wage war of conquest'.

Isn't this theory of environmental possibilism merely a preening Euro-centrism, extolling our familiar set of technical values over other options? Tasmanian Aborigines, stranded from the mainland 10,000 years ago, lost such early inventions as fish-hooks, needles, and even the ability to start fires. But is this really a loss, or just a cultural variant, a lifestyle choice neither better nor worse than our own neurotic, nuclear-threatened civilisation? Frankly, I find such adamant relativism perverse. In the broadest terms and across a 13,000-year canvas, it seems clear that some cultures happened to gain sway over others by carrying germs bred in their rich larders. That, at any rate, is a preliminary and plausible Darwinian answer—complex, filled with historically contingent detail—believable by civilised people who refuse to fall into reflex racism. It is the kind of explanation we shall need if we are to deal humanely with each other in a world where the scythe of mortality is replaced almost entirely by deliberate choice. We can no longer afford to demonise each other in the old, stupid ways, because what we stand to forfeit is not a few meagre years of life in an uncertain world but all of tomorrow's unfolding future.

SIMPLE QUESTIONS, COMPLEX ANSWERS

At the opposite extreme from such vast, global ecological and geographical vectors, underlying the DNA code and the proteins and cells which are its living expression, are the laws of chemistry and quantum physics. Those rules seem alien in their cool, reductive simplicity. How can life's rich diversity arise from less than 100 elements, from a handful of quarks, leptons and exchange particles, let alone the geometric forces of relativity? Can this wonder—the seething, fertile mind, the vast lighted darkness above us, the quantum flickering of radioactivity, the bounty of a million species—truly be nothing more than necessity catching the wing of chance?

One altogether new window has been widening for a decade and more upon the nature of things, and us: complexity. Chaos, the darling of science writers a decade ago, is the regime in which simple, deterministic rules generate wildly various outcomes, each depending on tiny differences in initial conditions. That unpredictable variation in outcomes placed its objects of study beyond the strict linear predictive tools of traditional science. Now chaos is better understood, and has come to be seen as just a small corner of an emerging paradigm: complex adaptive dynamic systems. Complexity theory promises a scrupulously rational, non-mystical model of the world in which 'emergence' itself, dismissed for generations as an illusion, re-emerges as a crucial feature. As simple elements interact locally, higher levels of global order and complexity spontaneously emerge, as if by magic. While quantum laws explain how to build a water molecule from hydrogen and oxygen, they don't account for simple emergent features like the twisty, sucking vortex swirling down a bath-hole.

Unlike magic, however, complexity can be mapped by tough

mathematics and subjected to scientific discipline. The Mecca of complexity theorists is the Santa Fe Institute, founded a decade ago by a bunch of elderly Nobel Prize laureates and wild young chaos experts from the nearby Los Alamos weapons establishment. The Institute was kickstarted by Murray Gell-Mann, inventor of the quark model of fundamental particles. The patterns Santa Fe has sought and occasionally found have been characterised superbly by Gell-Mann as 'surface complexity arising out of deep simplicity'.

Two of Santa Fe's faculty have risen to prominence among complexity fans: Chris Langton, whose 'artificial life' simulates organisms mutating in the mathematical space of a computer program, and Stuart Kauffman, a biochemist who showed that random networks can quickly stabilise into astonishing order. Kauffman's magisterial *The Origins of Order* may prove to share more than a mnemonic resemblance to Darwin's pivotal *On the Origin of Species* of 1859. Other Santa Fe visitors, such as Per Bak, find prodigious yet lawful complexity in a pile of sand or rice.

A biochemistry professor, Kauffman is interested in the ancient origin of life from non-living chemicals as well as the development of an individual from DNA code through embryo development to maturity. While Darwin and his Mendelian followers saw evolution as the result of competition for resources and fecundity between slightly variant organisms, Kauffman looks to other, deeper sources of order—not to replace Darwin's, but to amplify and accelerate them.

ORDER FOR FREE—MAYBE

Spontaneous order can be seen in oil droplets in water, whose molecules automatically form a sphere. Snowflakes have sixfold symmetry without benefit of natural selection. Proteins, coded by a linear string of DNA, only become effective when they spontaneously fold up into ornate lock-and-key shapes. There's nothing mystical about this, but Kauffman's great achievement has been to show how much order can emerge from linked networks of extremely simple elements. His favourite example is a web of 100,000 lights wired so that adjacent bulbs are switched on and off in a random cascade as they influence their immediate neighbours. In principle, we'd expect such a net to flicker through a cosmically vast set of possible patterns. In fact, ordered patterns spontaneously emerge, restricted to some 370 stable variants (about the square root of 100,000). It just happens that humans have roughly 250 distinct cell types and 100,000 genes, which suggests to Kauffman that these might well be the cell varieties generated, so to speak, by what he dubs spontaneous *anti-chaos*.

Santa Fe itself is generating a vision of life, mind and world (Gaia) self-adaptively seeking regimes at 'the edge of chaos'—the turbulent border between frozen order and noisy chaos. The paradigm's ambit is immense, drawing in theorists from economics, ecology, cell biology, computing, condensed-matter physics. Not surprisingly, it's still regarded with deep suspicion by the orthodox. The Sante Fe Institute, as science writer George Johnson has shown, stands quite literally at the intersection of several worlds that are almost invisible to each other. Los Alamos was for decades the centre of nuclear weapons and SDI ('Star Wars') research, and its supercomputers now host fertile work in non-linear dynamics, artificial life and other twenty-first-century growth areas. The Institute carries forward these new sciences of complexity, probing holes in classical physics and biology, making information itself the secret key to reality. At the same time, the starkly beautiful New Mexico landscape is home to Native American and Hispanic cultures to whom it is literally sacred. A hidden brotherhood, Hermanos Penitentes, flagellate themselves for Christ. Pilgrims trek to the village of Chimayo to daub themselves with miraculous healing dirt. Presumably they can't all be right about the nature of reality.[41]

Dig into the mind, Johnson suggests, and we fetch up at neural structures selected by evolution to find patterns in the world, inner as well as outer—even if the patterns are not really there. For Murray Gell-Mann, the purpose of the mind is to make data-compressions out of the abundant, flowing world. 'When you don't see compressions that are there, that is denial; when you see compressions that don't exist, that is superstition.'[42] Cultural blinkers plainly narrow our vision, permitting us to detect only some of the possible compressions. When women at last comprise half our scientists, we could conceivably find that much of today's science and technology is as quaintly idiosyncratic as the pueblo corn dances whose socially puissant magic is supposed to sustain reality.

Many of the complexity buffs at Sante Fe are convinced that by using very fast computers they already model aspects of the world that previous compression schemes can't reach. Without lapsing into the crystal-fondling New Age holism rife among many sun-dazed Anglo Santa Feans, they investigate 'collective phenomena, so-called emergent properties that can be described only at higher levels than those of the individual units'. Here, too, fashion holds sway.

CHAOS IN A HEAP OF RICE OR SAND

Science is not a remorseless intellectual engine strip-mining the universe, but a process of endless negotiation, two steps forward, one back and a leap sideways. A couple of years ago, when chaos theory was all the go, that neat phrase I used a moment ago sought to epitomise these discoveries: life and other complex phenomena were held to occur 'at the edge of chaos'. Life, it was supposed, seeks out this turbulent realm between frozen order and boiling noise and builds Gaudi cathedrals there. The spontaneous emergence of those complicated and lovely things is due to 'self-organised criticality'. Its emblem is a sandpile.

Or so Per Bak, a Sante Fe habitué, argues. His discoveries might have widely resonant implications. He has offered partial and startling new explanations for the frequencies of word use, the size and number of earthquakes, the odd computer program called the 'Game of Life', the extinction of the dinosaurs, the limits of planned economies, and why we will never get rid of traffic jams. And he thinks he knows how the brain works. It's all due to avalanches.

For Bak, the secret of life is a mathematical peculiarity called 'one-over-f ($1/f$) noise'. Unlike chaos, which has a boring 'white noise' signal in which one part is pretty much like every other, this particular regime creates a weird and charming pattern. Consider the scale of city sizes: for every metropolis of ten million inhabitants, there are ten with a million, a hundred with 100,000, and a thousand with only 10,000. It's the same with earthquakes, as measured by the Gutenberg-Richter law. Most are small, village-sized. Very few are dangerous monsters. Snow avalanches follow much the same pattern. Graph these frequencies and you get a spike on one side, tailing off quickly as the curve moves to the right. Map them on a logarithmic chart and you get a straight line.

Systems that are poised at self-organising criticality (SOC) are not too hot, so to speak, and not too cold. They bubble with novelty, but every now and then one of the avalanches of change will be triggered by some small, inoffensive event that cannot be identified in advance. It doesn't take a vast meteorite from space to wipe out the dinosaurs, Bak claims. SOC gives us catastrophe for free. It is the downside of complexity. Stuart Kauffman says of such ideas: 'I have a feeling that all this shit links together in some wonderful way.' I couldn't have put it better myself.

Or is this neat cartography flawed? Life—even 'artificial life'—keeps bursting out of the frames of our preferred compressions. 'Complexity', Peter Coveney and Roger Highfield have asserted, 'offers a means for transcending the materialistic limitations of reductionism and allows us instead to build a bridge between science and the human condition'. Perhaps. It is by no means clear to me, though, as I shall argue shortly, that 'reductionism' deserves to be a term of abuse. Embryologist Jack Cohen and mathematician Ian Stewart have offered another way to look at these issues. While complexity (our trillions of organised cells, say) often arises naturally from simplicity (our hundred thousand genes), one must also consider factors that Cohen and Stewart playfully dub *complicity* and *simplexity*.[43] Two copies of an identical genome, sent different signals from their surrounds, can grow utterly different bodies, far more divergent than twins raised in different countries. Equally, wildly different sequences of DNA can turn out creatures or organs you would have trouble telling apart.

While the universe of life and mind has evolved without what Dennett mocked as 'skyhooks', evolved by a drunkard's walk out of chaos, it has used a fabulous sequence of good tricks, forced moves, and (that other Dennettism) nifty 'cranes'. Its dazzling array of levels, from physics itself all the way up to human patterns of art, gossip, seduction and politics, are connected by level-crossing 'strange loops', a lovely and compelling term introduced by Dennett's sometime collaborator

Douglas Hofstadter. Organisation percolates up from below, but also sends its messages back down again as well as sideways, in complicity. The laws of nature, for Cohen and Stewart, are narratives honed 'until they capture very significant features of the way the universe works'.[44] Like life itself, really. Like us, the storytellers and listeners.

ARTIFICIAL LIFE, ARTIFICIAL DEATH

One remarkable outcome of such investigations is the quest for life forms simulated on a computer. Like complexity theory, artificial-life theory hints strongly that neither 'bottom up' nor 'top down' methods—neither the reductive nor the holistic—provide the whole truth. Self-replicating strings of numbers, dubbed 'genetic algorithms', can be set free to evolve in the virtual ecology of a machine. By sheer brute Darwinian power, their lineages yield powerful new programs designed by no human mind.[45]

Are these wonderful routines 'alive'? By some standards they have more right to the word than viruses do. They are the very contrary of de Man's deconstructive beliefs about language: arbitrary yet lawful, coherent, combative, striving to declare their selfhood in a language of simulated DNA that speaks to us all too clearly. Perhaps that is because their kind of life, or pseudolife, is close to the kind we know most intimately, from within: the life of mind. But what is mind? Does it float free of the brain, a kind of non-physical ectoplasm? (And if so, would artificial life systems generate a kind of eerie cyberplasm?) Is it doomed to disintegration with the failing of the body, the brain? Most piercingly: can it be sustained indefinitely by an immortal generation of rejuvenated humans, or moved from the protein body to a newer, more enduring substrate? Let us turn now to those ancient, yet astonishingly renewed, ceaselessly perplexing questions.

four: mind – the machine in the ghost

Multiple, converging lines of evidence impelled me to propose that human beings have evolved as a species to possess at least seven distinct forms of intelligence— defined as the ability to solve problems or fashion products that are valued in at least one cultural setting or community. My initial list of intelligences included linguistic and logical ... spatial ... musical ... bodily kinesthetic ... and two forms of personal intelligence—one oriented toward the understanding of other persons, the other toward an understanding of oneself. Recently, I have enlarged my list to include an eighth ... the apprehension of the natural world ...

Howard Gardner, Professor of Education, Harvard Graduate School of Education, in *Extraordinary Minds* (1997)[1]

The proper study of mindkind, as the eighteenth-century poet Alexander Pope didn't quite say, is mind.

That isn't a slur on the human body, on our passions, intuitions, blood-thrilling joys and sorrows. Mind, or consciousness, just *is* the body experiencing its rich throng of impressions and thoughts, recalling the past, relishing the present and preparing the future. Thus, at any rate, runs the consensus of today's experts in the cognitive and neurosciences. Not everyone agrees. Many religions teach that each human is a kind of metaphysical Siamese twin, brute matter yoked to sublime spirit. Our suffering atoms might disperse upon death, but soul, mind, spirit (not synonymous, but linked in their ineffability) persist and find justice or at least peace.

While there's no scientific evidence for this claim, adherents see the strongest proof in experience itself. We know what we are, from the inside. In the language of philosophers, our most humble experience is drenched in *qualia*—that is, the intimate subjective *feelings* or *qualities* that no Positron Emission Tomography scan will ever detect: the sweet taste of a ripe peach, the glowing, calming redness of a sunset (not at all the same thing as the energy spectrum of its light). Qualia, it seems, are the very habitation of mind. If so, how can mind be nothing more than the brainy body in action? On the other hand, cognitive-science researchers and evolutionary biologists are gradually showing how mind can emerge unmysteriously from complex neurological structures. Plenty of gaps remain in the chains of explanation, but arguably the story is coming together. Intelligences, awareness and even consciousness have developed—and still function—by Darwinian principles.

Is Consciousness a Hard Problem?

Some, as I say, disagree strongly with this estimate. A professor of philosophy in Santa Cruz, David Chalmers, is an Australian who started in mathematics and computing at Adelaide and Oxford before jumping in at the deep end in Indiana, with computational guru Douglas Hofstadter, to explore mind by traditional and scientific means. For several years he has been at the eye of a storm because of his forthright claim that consciousness is what he calls a 'hard problem', that mind has its own special laws, even though it arises from (or 'supervenes upon') matter.

The celebrated and feisty philosopher John R. Searle, himself a foe of reductive materialism, recently engaged Chalmers in the *New York Review of Books*, declaring his assertions absurd.[2] Chalmers responded vigorously, and the debate continues. Perhaps the most startling aspect of this skirmish is that Chalmers isn't a desiccated Jesuit or a bald pipe-smoker in a tweed jacket, but a strikingly good-looking young man with flowing heavy-metal hair. Meanwhile, philosopher Daniel Dennett bluntly denies that we *are* intrinsically different from other kinds of life, even if mind itself is rare. While viruses and bacteria are undeniably mindless robots bustling in our flesh, we must not, Dennett warns, 'take comfort in the thought that they are alien invaders, so unlike the more congenial tissues that make up us'. Look at them under a powerful microscope and we see beyond argument that the brain's billions of neurons are just cells scarcely different from those of germs or the yeast cells in beer vats and bread dough. 'Each cell—a tiny agent that can perform a limited number of tasks—is about as mindless as a virus.'

Where, then, does mind comes from? From the organisation, the networks, the nuanced dance of these specialised ninnies. Not from a mysterious soul, but from the baroque layout of the cellular arrays. Does this mean that dogs also have minds? After all, their brains and nervous systems are not really so much more primitive than ours. And certainly there seems to be someone home there, behind the brown loving eyes. Or is that an illusion, projected by the human spirit (which is what we call our wonderful, vastly complex machinery)? But if dogs, what of cockroaches? Ants? Rocks? Dogs, Dennett admits, might be a special case, for they have co-evolved to resemble us in many respects.

DARWIN MACHINES

William H. Calvin, a Seattle neurophysiologist and polymath writer on science, argues that consciousness—or intelligent awareness, at least—arises because brains are 'Darwin machines'.[3] Stimuli from the outside world (or from memory) trigger specialised parts of our brains into cloning temporary representations of things we've seen or thought before. The cloned activation patterns multiply in hexagonal arrays of cells half a millimetre apart, hungry for brain-space, synchronising, reinforcing or inhibiting each other. Each is a kind of fragment of a thought or recognition. Calvin's suggestion is astonishing, testable, and links minds to the rest of the evolutionary universe. Meanwhile, Chalmers is unconvinced. One can imagine a world of zombies, he asserts, just like ours but *lacking* consciousness. The brains of these zombies mimic ours precisely, but there's no light on inside. They act and speak and laugh and 'love', but are mere automatons. Since this nightmare is logically possible, Chalmers says, consciousness must be something over and above mere neural structure in action.

What is it, then? Information, he concludes. Not an answer to give the faithful any comfort, admittedly. Chalmers believes a complex artificial intelligence system would be conscious. So he is not proposing to reinstate an immaterial soul. He puts it neatly: 'Experience is information from the inside; physics is information from the outside.'⁴ Inside *what*, though? Inside the mind, with the qualia. But that leaves us where we came in. Besides, I would argue that we cannot truly imagine a zombie world, any more than we can truly imagine a world exactly like ours, full of jittering molecules *but without heat*. Stewart and Cohen provide an amusing analogy to support this suspicion. Imagine a *zombike*, they suggest, 'which is *exactly like a bicycle in every way* except that it does not move when the pedals are pushed. Oh, mystic miracle of ineffable immateriality, the source of motion in a bicycle is not anything physical!'⁵ The zombie analogy, at root, is no more persuasive nor even intelligible.

NEW WINDOWS ON THE MIND

During the last half-century, cognitive science has rewritten our understanding of the mind. Tom Wolfe declared recently in *Forbes* business magazine, of all places, 'Neuroscience ... is on the threshold of a unified theory that will have an impact as powerful as that of Darwinism a hundred years ago.' We live in an age, Wolfe noted ruefully, in which it is both impossible and pointless to avert our eyes from the truth revealed by the subtle new brain-scanning technologies. Even so, somewhat to its embarrassment, brain science and computational disciplines are still unable to create an artificial intelligence. By a piquant if somewhat Frankensteinian paradox, the first step might turn out to be uploads—human minds scanned and copied into machine memory, residing in a blend of virtual and biological realities. Before anyone can hazard such a bold step, of course, we need to test how ordinary minds work (human and otherwise)—indeed, learn *what they are*.

Let us glance, modestly, inside the human mind (pre-immortality version). Or do I, after all, mean 'brain'? Here's a good starting point: *minds are the way the brain minds the body* (including the brain itself).

Despite the traditional aura of mystery and reverence, the human mind can seem a vulgar instrument, or at least mine does all too often.

It's a silly thing, avid for the slightest punning chance to torment one word into another so I can fall about laughing. And the indecency is only worsened by the pitch of abstraction at which it occasionally vibrates. The wittiest joke I ever heard, the one which most swiftly cast me from noble musing to a horrible state of shouting admiration, was also the most ridiculously donnish. But first—

Have you heard the story told by British humorist Frank Muir (or it might have been his colleague Dennis Norden) about a humiliating last-minute invitation to a fancy-dress affair on the theme of Tragedy? The poor fellow racked his brains for a suitable outfit, but time was short. Inspiration saved him. Norden (or perhaps it was Muir) presented himself in dinner suit and bow tie, topped by a novelty beanie with a spinning battery-operated propeller. His hostess was aghast, herself all bloody as Clytemnestra on a country lawn of Lears, Hamlets, and a bloody-socketed Oedipus. 'That's not tragic!' she cried in rebuke, pointing with trembling finger at his slowly rotating headgear. 'True enough,' mumbled the abashed Muir (or it might have been Norden), 'but you must admit it's moving.'

WORDS AND THINKS

The power of verbal categories over our illusory sense of direct, unimpeachable perception of the world is known in linguistics as the Sapir-Whorf hypothesis, now somewhat in eclipse. Learning of this wonderful conjecture was one of the high points of my adolescence. My God! I thought, dazed. Does this mean that the world is radically mutable, and by so simple a thing as *language*? Our inner world was shaped by the words available to our use! Some years later I mentioned this theory to an urbane academic friend. 'Whorf's hypothesis?' he muttered, hardly impressed. ' "Ontology recapitulates philology".' That's the terrific joke, by the way. You hate it. Oh well. I fell off my chair and had to be calmed. Let me explain.

I couldn't believe I'd heard him correctly, which happens a lot when you entertain Benjamin Lee Whorf's hypothesis. I supposed that he'd said (though I couldn't imagine why), 'Ontogeny recapitulates phylogeny.' That is the spurious nineteenth-century Biogenetic Law promoted by Ernst Haeckel to account for the way embryos seem to pass successively through the stages of their evolutionary history. And then, in a mental flash of what Arthur Koestler called *bisociation*—slamming together two disparate frames of reference—I was hooting in purest delight, like a loon. Ontology, the science of what exists, reprises or echoes philology, the science of *utterance*. My uncontrolled laughter is certainly strange when you consider how extremely *remote* the jest had been from, say, the sight of a fat toff busting his ass on a banana peel.

I had been surprised, you see, by my own mind.

How is that possible? Am I not my mind? Is my mind not I?

Of course I am my mind, and the body it's running (that's running it), but I'm not privy to all its contents and operations. What startled me was an insight forged inside a partitioned-off and inaccessible (or 'unconscious') part of it, a pun-making and pun-recognising gadget that presumably sits there poised like a mousetrap in the folds of the linguistic cortex on the left side of my cerebrum, waiting for a chance to snap shut on some wriggling word.

After all, how do we think up whimsical gags? We prime ourselves, get in the right mood (whatever that is), and out comes … something funny. Mildly funny or hilarious, witty or worth a groan. Either way, *new*! Amazing, really, in a universe of machines.

Sitting at my kitchen table reading the other day, I heard Sir John Gielgud speaking on the radio as Richard III: 'Now is the winter of our discontent made glorious summer … ' I got up and made a pot of tea, idly wondering, 'How would Shakespeare have written that if he'd been an Australian?' I played with the sounds, disconnecting the meaning. Let's see, *Noosa*, a sunny vacation spot favoured by penniless hippies prepared to live rough. 'Noosa,' I mumbled, 'winter arvo, discount tent.' I smiled, but it was too fragmented, too *designed*. Besides, one would be obliged to explain and footnote, which is death to jokes, to explain that 'arvo' is Aussie vernacular for 'afternoon'. What about an FM-music station version: 'Wow! Here's the winner of our disk contest!' Not too good, since it's not relevant to anything except the raw sounds of the original phrase. I sipped my black tea, wishing it were something stronger. An unusual Australian word nuzzled into my conscious:

'wowser', a strange local term for the purse-lipped, abstemious prohi-
bitionist who is the exact opposite of a wine-seller. Which led me
instantly, and with no further conscious effort, to the genuinely funny
'Wowser: the vintner of our discontent.'

I, like you, have a complex set of schemata—interpretation
machines—buzzing away in various not-quite-conscious regions of my
brain/mind, shouting for attention. They generate structures and
oppositions to those structures, and throw in forgotten quotes, all that
routine mental business. *I think I think, therefore I think I am.* A writer's best
phrases come out of inaccessible regions of the mental machinery. One
that surprised and pleased me the other day was 'quantum quackpots
and black holists'. The first phrase was a fairly ordinary punning rear-
rangement, a little heavy-handed, even arch, but the second leapt unbid-
den from my typing fingertips and left me sniggering for a moment or
two. It did not come from some mysterious 'collective unconscious', as
a Jungian might guess. I thought it up myself! Still, the 'I' who thought
it up was *merely a small part of me*, and only let the rest know of its presence
by typing the words onto the screen ...

THE WORD-PROCESSOR THEORY OF MIND

And that's a prime clue, believe it or not, to the nature of brain and mind
alike. Daniel Dennett has encapsulated this organising principle in a
prosaic desktop computer phrase: he favours a 'multiple drafts theory of
consciousness'. Mind is not an angel tragically trapped in flesh, but a text-
processing program. *MindPerfect 5.1* for DOS, very nearly obsolete but still
widely in use. *Mind 6.0* for Macs, What You Think Is What You Get.
HTML hypertext for the World Wide Mind. That kind of thing. Of think.

To tell the truth, I'm in two minds about Dennett's claims to have
solved the mystery. Or do I mean two minds are in me? Maybe more than
just two minds are scurrying about inside my head, since I seem to be quite
unsettled in my opinions. Yet 'opinions' aren't the same as 'minds', are
they? Or am I confusing 'minds' with 'selves'? But 'selves' have 'bodies'
as well as 'minds', don't they? So should that be 'are' bodies and minds,
rather than 'have'? Or is that true only until the body dies?

Dennett's playful philosophical method, which I'm caricaturing, makes a delightful contrast to the preening obscurities of its most notable rival doctrines, Heideggerian phenomenology and its Anglo-French heirs. He and his materialist colleagues have been beavering away at the problems of the contents of consciousness, will and intentionality since the mid-1960s, updating their formal arguments in line with floods of research issuing not just from the philosophy seminar but more importantly from the scientist's bench and the psychologist's test room. With the brilliantly agile computer scientist Douglas Hofstadter, Dennett co-edited an accessible anthology on the topic, *The Mind's I*, drawing upon fiction and artificial intelligence research as well as Anglo-American epistemology (the study of how we know things). His approach to the secrets of the self is fairly summarised in the charming but quite serious gag by Hofstadter, author of the formidably interdisciplinary *Gödel, Escher, Bach*: 'Is the soul more than the hum of its parts?'

Like a computer humming happily to itself, accepting information from the outside world and processing it within a suite of programs that in turn derives from that outside world, does the brain 'run the software' of the mind? Or is this functionalist formula just the most modish in a line of opportunistic analogies, a desperate but doomed attempt to shoehorn the ineffable into whatever current box of tricks (clockwork, steam engine, electronic gadget, chaos theory) is borrowed by eager and shameless reductionists? That's an objection risen again into philosophical view due to the efforts of enthusiasts like Chalmers for the 'hard problem'. What exactly can consciousness be? What are experienced qualities, in a world where, science assures us, nothing but matter and energy and perhaps information truly exist? It's a crucial question that will be met head-on (so to speak) in the early and mid-twenty-first century, as we begin to deal with issues like nanotechnological reconstruction of the ageing or even cryo-preserved body, and then the attempted uploading, or copying, of human minds into 'machine platforms'.

THE DUALIST ILLUSION

René Descartes, three and a half centuries ago, decided that the world was a machine, with one exception: the human mind or soul. His meditations led him to conclude that this non-material entity must be anchored somehow to the rude physical flesh, presumably somewhere in the head. The pineal gland, neatly in the middle of the brain, seemed a good candidate for mind-body docking. Nobody takes this pineal fancy seriously today, but an underlying Cartesian dualism remains surprisingly persistent. It's supported by an illusion we almost inevitably fall into, that there's an 'inner person' behind the eyes, somehow immune to age, continuous despite bodily changes, morally responsible even when moods sway us (for moods seem more 'physical' than trains of thought, don't you think?).

Dennett has dubbed this illusion the 'Cartesian Theatre'. With a barrage of superbly chosen clinical and logical examples, he has repeatedly demolished its claims on our intuition. Oliver Sacks (whose most celebrated patient mistook his wife for a hat) helped prepare the ground for us to accept this sort of argument. Lay readers enjoy Sacks' beguiling case histories; he is, you might say, the neurologist from the Lake Wobegon Hospital for the Brain-Buggered. Sacks shows repeatedly that almost any given aspect of what we consider integral to selfhood can be damaged with great precision when brain or bodily function is hurt. The soul's hum falters. An astounding report of this kind is from a 1983 Munich clinical paper on a woman who is *motion-blind*! She finds pouring tea tricky, because 'the fluid appeared to be frozen, like a glacier'. Crossing the street is scary: 'When I'm looking at the car first, it seems far away. When I want to cross the road, suddenly the car is very near.'[6] Her brain edits out the intervening motion, turning ordinary experience into a lethal video clip. And yet the dignity of the human person need not be compromised, let alone denied, in this kind of 'debunking' demonstration. Sacks' wounded patients break our hearts, but we are left with insight rather than a gush of sentimentality.

If we remove the homunculus (or shrunken inner observer) from the imaginary cerebral theatre, what are we left with? Dennett's word-processing model proposes mind as multiple drafts of a document generated, edited and acted upon in a massively distributed parallel computer. This seems at once vulgar (mind as sitcom script-writing conference) and ineffective (for surely an inner consciousness is still needed to 'read' the 'document'?). Dennett's triumph has been to lead sceptics by small stages, each modifying yet adding to its predecessors—like a series of drafts of the argument, you might say—until we confront counter-intuitive conclusions that seem altogether convincing.

Is such facile ease a reason in itself to distrust a theory that, like a magical trick, defies our expectations? 'I don't view it as ominous,' Dennett declares, 'that my theory seems at first to be strongly at odds with common wisdom ... The mysteries of the mind have been around for so long, and we have made so little progress on them, that the likelihood is high that some things we all agree to be obvious are just not so.'[7]

SEEING THINGS BEFORE YOU SEE THEM

It is difficult even to hint at the huge amount of evidence and argument cognitive scientists and philosophers have marshalled lately in support of this account. Ranging from very peculiar sensory illusions routinely evoked in the psych lab, all the way to pathological 'multiple person-alities' found in the victims of horrendous child abuse, they cut the ground from underneath any intuition we retain that mind and self are clearly unitary and transparent to inspection.

Perhaps the most striking example is an experiment in which a red and a green spot of light flash in rapid alternation against a dark back-ground. How would you expect to see this sequence? It's the principle on which television is based, after all. When we sit in our living rooms, watching the small screen, coloured phosphor dots flicker, weaving a sequence of static pictures that our minds blend into the illusion of people in motion. Still, this red/flick/green lab experiment is very weird. The red dot looks as if it's moving across the screen, then abruptly switches to green in the middle.[8]

Think about it. The damned thing changes colour *before* your con-sciousness is aware of the green spot! The Cartesian Theatre model tells us that the inner watcher must have been prevented from seeing this sequence until the whole thing was over, whereupon a kind of retro-spective censorship or rewriting of history got foisted on us. Dennett argues lucidly that this is the wrong story to tell. In fact, the observer's point of view is dispersed, 'spatiotemporally smeared all over the brain'. There's no single inner watcher to be fooled. We are composites, spread out slightly in time and space. Perception is not merely prone to illusion; it's based on it.

Mind is a virtual machine, built by culture, running on the brain's neurological parallel processing system. This is a holistic theory, of sorts, but with quite specific mechanisms and predictions. Perhaps Dennett's strangest suggestion, to those interested in artificial intelligence (AI) and the future prospect of uploading a living mind into a machine implementation, is that the mind's 'virtual machinery' uses a linear architecture like today's supercomputers, though it is obliged to run on the brain's quite unsuitable machinery. It has become fashionable to suppose the reverse—that 'connectionist' or 'neural net' data processors are the wave of the future, because they more closely resemble the brain.

Beyond such specialist arguments, though, is the question that keeps me in two minds about Dennett's claims to have solved consciousness. Has he, after all? Don't these multiple drafts still feel from the inside like a unified self? Don't other people present themselves with a certain fundamental unity? Of course, admits Dennett. It's because humans have evolved as storytellers, just as beavers have evolved as dam-builders. A side effect of grammar is 'to (try to) posit a unified agent' speaking these words, 'to posit a *center of narrative gravity*'.[9] Just as a body's mass is dispersed throughout its volume, but its motion is governed by an imaginary point, its 'centre of gravity', so too with us as *experiencing persons*. 'We build up a defining story about ourselves, organized around a sort of basic blip of self-representation', like the blip on a radar screen. 'The blip isn't a self, of course; it's a *representation* of a self'.[10]

The hilarious thing about this hot new cognitive science story, to me, is that it closely mimics the account evolved in recent years by post-structuralist philosophy, semiotics and psychoanalysis. Those uptown intellectual boulevardiers—with their difficult and frequently derided jargon of 'antihumanism', 'subject positions', 'discourse formations', 'deconstruction' and 'dissemination'—turn out to have an eerie resemblance to Dennett's down-home empiricists. Dennett once acknowledged this in a nicely comic moment, citing (with 'mixed emotions') a version of his theory he found in David Lodge's campus parody *Nice Work*, attributed to a fashionable English Department deconstructor.

Significantly, Dennett's project has been hailed by Professor Richard Rorty, perhaps the leading philosopher who is at home in both these divorced traditions. So while there is no doubt that Dennett's version of self is just one draft in an evolving discussion, it has every prospect of remaining at the debate's centre of gravity.

WHOLES AND PARTS

The balance between a kind of holism ordained by any massively distributed theory of the brain, and the reductionism required to investigate its properties at close range, bedevils any discussion of these issues. We shall find ourselves returning to this background dispute, this nagging tooth of methodology. Consider the case of the late Lewis Thomas, a gifted essayist and former director of the famous Sloan-Kettering Cancer Center in New York. Thomas wrote eloquently for 20 years in the prestigious *New England Journal of Medicine* about whatever took his fancy, which was rather a lot: medicine, of course, both bedside and high-tech; immunology, cancer research and cell biology especially; ecology; the roots of words. His first collection, *The Lives of a Cell*, bowled me over in the 1970s. Musical and learned, gentle-voiced and tough-minded, Thomas craved the global generosity of holistic approaches while admitting candidly that uncompromising reductionism remains the successful method of choice in science.

A recurrent theme in Thomas' essays was biologist Lynn Margulis' wonderful argument (which we have seen in a previous chapter) that the complex cells we are made of are themselves colonies. The mitochondria which power life have their own DNA, passed down only through the maternal line, and seem to be relics of a symbiotic partnership with larger protocells when the world was much younger. So too, perhaps, Thomas mused, might we stand within the economy of Gaia, our planet viewed as a self-sustaining quasi-organism. Here he ventured close to gibberish, in my view, but it might (alas) be the kind of gibberish we need to get us through the stresses of the twenty-first century:

> *Given brains all over the place, all engaged in thought, and given the living*
> *mass of the earth and its atmosphere, there must be something like a mind at*

work, adrift somewhere around or over or within the mass ... It is, if it exists, the result of the earth's life, not at all the cause. What does it do? It contemplates, that's what it does.[11]

No hypothesis could be more repugnant to Nicholas Humphrey, a lucid research psychologist and colleague of Dennett. Strictly materialist, he once accepted a fellowship funded to study psychical phenomena—so as to learn why otherwise rational people hold to such doctrines as post-mortem survival. A disembodied mind is, for Humphrey, a contradiction in terms, since the mind is exactly the activity of a brain in a body built by evolution to deal with the teeming reality of the physical world. Like Dennett and Chalmers, he pursues the toughest of all assignments, a satisfying account of consciousness. For even if the structure of neurons and chemical transmitters is clarified, even if PET scans literally show a thought passing through a brain, how can a sack of atoms add up to feelings, to sensations and awareness?

His answer is simple enough: consciousness is sensation—not 'perceptions, images, thoughts, beliefs'—become self-aware in feedback loops inside the brain. (Hence he insists, unlike Dennett, that robots will never attain consciousness, since they will lack the evolutionary history of our senses.) Sights feel different from sounds and touches due to an evolutionary shaping process, because we need to tell these channels apart in an instant, for our survival. Feelings are '*activities* that we ourselves engender and participate in—activities that loop back on themselves to create the thick moment of the subjective present.'[12] Like all the arguments we are glancing at, Humphrey's rich case can hardly be summarised so curtly. Still, while I sense that it's persuasive enough, I perceive, think and believe that, in exalting sensation over the rest, he may be too hastily in flight from the reigning computational accounts of mind.

ARE WE INTRINSICALLY BETTER THAN MACHINES?

Perhaps we too easily assume that people are utterly different from machines, even today's trendy neural networks. Can't we do something no machine ever displays: break free of set patterns, act creatively? Margaret Boden, professor of psychology and philosophy at the University of Sussex, denies this. Deft with musicology, romantic poetry and the history of science as she is with computational psychology, Boden gives such complacent prejudice short shrift. Creativity, she observes, implies a capacity to jump free from a set of constraints into a surprising 'impossible' solution. To a modest extent, computer programs have already done this. They have rediscovered laws of physics using the rude data and heuristics available to scientists in centuries past. Computer programs have written passable jazz (though not convincing poetry), and found unexpected solutions to mathematical puzzles which, if uttered by a bright young student, would earn an elephant stamp.

Here's a simple example (too smug, too hasty, I got it wrong): if ABC changes to ABD, what will MRRJJJ become? What does XYZ turn into? A smart program named Copycat has provided a range of solutions. Answers: since ABC maps to 123, ABD equals 124; hence JJJ becomes JJJJ, yielding MRRJJJJ (not, say, MRRJJD or MRRKKK, let alone MRRKKKK). And XYZ: XYZZ? XYY? XYA? Maybe. But clever Copycat saw that A maps onto Z. Running backwards, we get WYZ. (Elegant. Think about it.) Is that artificial mind creative, or what?

I once wondered idly if our fascination for black holes in space might include a kind of accidental hard-wired neurological explanation. A recurrent nightmare of mine as a small child was both terrifying and non-representational; I have always assumed it tapped into some primary perceptual building-block schema, the kind of cognitive template we construct our subjectivity from. In my nightmare, I was in the presence of unavoidable shapes like infinitely elongated cones, black and awful, curving downward sharply from a flattened top. They looked, I realised much later, like embedding diagrams of black holes in hyperspace ...

Cognitive science is often regarded by humanities specialists as a playing field for deluded positivists whose search for computerised artificial intelligence, doomed in advance, proves how *derrière-garde* their gung-ho enterprise must be. A sceptical estimate of cognitive science is not altogether unjustified. Francis Crick, co-discoverer of DNA's structure and nowadays a brain researcher, cites a cautionary quote from John Searle: 'As recently as a few years ago, if one raised the subject of consciousness in cognitive science discussions, it was generally regarded as a form of bad taste, and graduate students, who are always attuned to the social mores of their discipline, would roll their eyes at the ceiling and assume expressions of mild disgust'.[13]

TOP-DOWN, BOTTOM-UP

Neuroscience is typically done 'bottom-up', following the assumption that big things are built from little things, and take much of their form and function from the constraints imposed at the lower levels of organisation. On the other hand, there has been a huge shift in emphasis during the last decade or so. Plenty of folks working in the neuroscience labs, as noted earlier, have realised that 'complexity' and 'emergence' are the watchwords for their developing understanding of self and its gooey components under the skull. Although I defend the computational model, the best explanations will transcend metaphors based on rigid styles of programming in which information is sent through a processing unit in a stream of well-defined tasks. John McCrone observes, by contrast, that

> modern neuroscience sees the brain as a dynamic neural network. A 'state of information' has to grow organically, evolving under the pressures of positive and negative feedback until it reaches a state of balanced tension. In such a network, it is not the speed of traffic along the individual wires that counts but the performance of the entire network as it settles into a 'solution state'.[14]

This does not mean we must revert to the Cartesian Theatre, that mythical site deep within the brain (or outside it, in the 'soul'). Rather, it implies that thought and awareness resemble the self-organising—or, better, *mutually* organising—patterns created by the solar system as its planets and asteroids and comets orbit our central sun, falling into typical 'attractors' or balanced orbits while remaining open to perturbation, especially from the changing geometry they create with respect to each other. No machine calculates these dynamical patterns, they simply emerge from the physics of space and time. Yet we can *mimic* those patterns by cunning and insightful equations that evolve with lightning speed inside a computer. Perhaps the way the brain works is like that: it just settles into place after a disturbance, rather as Per Bak's sand piles up into a cone when it is trickled from above. No computation is needed. And out of these astonishingly ornate and hidden orbits about the attractors of the mind, a fairly continuous 'self' is contrived.

BUILDING A KITTEN

Somewhere between the two paradigms is 'evolvable hardware', an approach to AI pioneered by Hugo de Garis, an astonishingly arrogant Australian-Briton (and 'masculinist' who comically declares genius to be a male prerogative). De Garis heads the Brain Builder Group (BBG) of the Evolutionary Systems Department, ATR Human Information Processing Research Laboratories in Kansai Science City, Kyoto, Japan.[15] A former theoretical physicist who studied under maverick quantum theorist David Bohm before turning to AI, he looks toward the era of what he calls Avogadro Machines, or 'artilects', artificial brains with a trillion trillion components, rivalling human brains.

In a homogeneous Avogadro Machine, a vast number of copies of a small variety of simple components interconnect in simple top-down ways, so its functioning remains simple enough for us to track and understand. A Heterogeneous Avogadro Machine, with huge numbers of complex components connected in complex ways, would surpass human understanding. While heterogeneous machines can't be designed from the top down, they *can* be built by evolutionary techniques. Those would evolve cellular automata machines (CAMs), neural net modules able to perform a range of tasks.

De Garis and his team have already put ten million artificial neurons into their CAM, capable of updating 200 million CA cells per second, and look toward programmable hardware in a CBM (CAM-Brain Machine) with a trillion refreshes per second. In April 1998, they were confident of putting a billion neurons into an artificial brain by 2001, and announced that they soon expected to see

the completion of two parallel tasks, namely the design and fabrication of the CBM (which was started in January 1997, and should be completed by the summer of 1998), and the creation of a 10,000 module artificial brain architecture to control a robot kitten's many behaviors. Starting in February

1998, the construction of a robot kitten called 'ROBOKONEKO' (in Japanese) has begun, and should be completed by April 1999. The 10,000 modules for the artificial brain will be evolved in 1998 and 1999 with the CBM, and placed in the large RAM memory. The [...] RAM brain will link via radio antenna with the (life sized) kitten robot. After 2000, the BBG hopes to work on more ambitious projects, such as household cleaner robots, and with substantially more brain builder researchers on the team.[16]

Is it sensible to speak of such early AI machines as 'artificial brains'? However powerful their modular structure grows, however advanced beyond the robotic kitten stage (already a quite staggering achievement), will they really have the capacity to host an artificial mind, or even an artificial self? But this is where we came in.

WHAT *IS* A SELF, ANYWAY?

A recent academic conference on mythopoeic literature was entitled 'Desperately Seeking Selfhood'. While the room had its share of pre-deconstructive academics and Jungians whose notions of wild transgression seemed confined to fairytales and *Star Trek*, many papers tore at certainties of identity, of self. Nor was this accidental. Far from *seeking* selfhood, the main projects in both humanities and life sciences today desperately *flee* from it.

Or so it might seem. In fact, that's chiefly true of the poststructural insistence on the decentred and disseminated self, the doctrine of self as unauthored text taught by French theorists Jacques Lacan, Roland Barthes, Jacques Derrida and Julia Kristeva. Self, they argued, was largely 'written' into existence by impersonal culture, rewritten by every reader—every other human—encountering it. (But *who* are these readers—for their selves, too, must be just as volatile and uncertain?)

Cognitive science holds the patents on most of the mind paradigms in play outside the humanities departments, and it in turn, as we have seen, is slowly picking apart the self into a vast flow-chart of dedicated modules, of specialised mental organs and their distinctive tunes. Oddly, unlike poststructural accounts of the self and its representations of body and world (the kind Daniel Dennett found so reminiscent of his own work), this new story is finding a certain fondness for individuality, systematic emergence, an integrated 'self' implicit in each unique genetic program.

Chris Langton is the brilliant oddball who almost single-handedly invented what has come to be called A-life or 'artificial life': computerised simulations of living processes. Hang-gliding some two decades ago, he botched his landing and smashed into the ground, breaking most of his bones and badly damaging his head. His description of what it felt like to come back on-line as a person is extremely provocative:

> *I had this weird experience of watching my mind come back ... I could see myself as this passive observer back there somewhere. And there were all these things happening in my head that were disconnected from my consciousness. It was very reminiscent of virtual machines ... I could see these disconnected patterns self-organise, come together, and merge with me in some way ... It was as if you took an ant colony and tore it up, and then watched the ants come back together, reorganise, and rebuild the colony.*[17]

These are suggestions rather more rigorous and at least as interesting and fecund as, say, those of the poststructuralist psychoanalyst Jacques Lacan (still fashionable although he died in 1981, and his key work had been finished long before that). Langton's detailed account of his mind rebooting is profoundly suggestive to anyone toying with the notion that selfhood is comprised of multiple subsidiary 'minds', intelligences, modules and agents, mostly inaccessible to consciousness. But we can't help asking: who was the *me* with whom these tiptoeing mental fragments of the old Langton were merging? Was it just the sketch of the '*full me*' the rest of us have, or, more accurately, *are*? Or was his 'self', and ours, a sort of memory-limited buffer, or working space, where subordinate modules might dump their partial comments about the world, to be written together there into the text of the experienced self?

'So my mind was rebuilding itself in this remarkable way,' Langton recalled later.

> *And yet, still, there were a number of points along the way when I could tell I wasn't what I used to be, mentally. There were things missing—though I*

couldn't say what was missing. It was like a computer booting up. I could feel different levels of my operating system building up, each one with more capability than the last. I'd wake up one morning and, like an electric shock almost, I'd sort of shake my head and suddenly I'd be on some higher plateau. I'd think, 'Boy, I'm back!' Then I'd realise I wasn't really quite back. And then at some random point in the future, I'd go through another of those, and—am I back yet or not? ... When you're at one level, you don't know what's at a higher level.[18]

Now of course, this report can't be taken, at face value, as proof that the mind is discontinuous or modular, let alone that it is organised like a linear computer program, using some kind of familiar systems architecture built out of suites of subprograms, able to call specialised subroutines. In fact, we're pretty certain that the human mind *isn't* very much like that at all. If it has a computational basis, its architecture will be massively *parallel* rather than *linear*, despite Dennett's claim that we function in a linear if weirdly distributed fashion. What's more, Chris Langton was on industrial doses of drugs, so their impact, coupled with his own interest in computation, might have produced a confabulated construct in his damaged brain, no more veridical than the strange experiences reported by people who are certain that they've been abducted by UFO occupants.

Nevertheless, Langton's testimony fits in very evocatively with other material that has surfaced from neurological case studies and neuroscience investigations of how both *normal* and *abnormal* brains appear to function.

A Kinder, Gentler Materialism

Happily, there's been a shift in emphasis during the last decade or so. This tempers the stern materialism of Francis Crick's so-called 'Astonishing Hypothesis': namely, 'that each of us is the behaviour of a vast, interacting set of neurons', and—culture aside—nothing more. Crick has deliberately restricted his own post-DNA research to the problem of vision, which is more tractable than seeking wildly for a general theory of mind. What is the path, though, from the retina to our aware sensation of *seeing*? Ray Jackendoff, a linguist influenced by Noam Chomsky of MIT, has suggested that consciousness dwells neither at some integrated level of the entire cortex nor at the lowest level of stupid individual neurons. The MIT philosopher Jerry Fodor, whose important texts include *The Modularity of Mind* and *Psychosemantics*, insists that our (surmised) central interpreters are *unlike* modules—being dispersed and non-localised, in some way mysterious and holistic. Jackendoff is convinced that even the inner interpreters are partitioned off.

How is it, then, that we coordinate our representations of world and self? We do so through abstract mental models. These are never available to awareness, but we can be sure they must exist as the templates for our neural computations. We never *see* things in true three-dimensional form, for example, since that would require us to observe, in the same instant, the back and front and top and bottom and insides of an object. But we certainly construct *interior models* of the world with this rich 3-D character, and it's in these abstract models that different sensory modalities—sight and touch, say—are brought into common registration.

Smelling Noises

Strange tales in support of this view pop up in neurological studies of both normal and abnormal brains. Fans of nineteenth-century poetic symbolism and impressionist painting will recognise 'synaesthesia', a curious condition where one sensory channel seems to be cross-wired to another. Forty years ago, novelist Alfred Bester's vivid protagonist Gully Foyle was slammed into synaesthetic confusion in *Tiger! Tiger!*

> *Touch was taste to him ... the feel of wood was acrid and chalky in his mouth, metal was salt, stone tasted sour-sweet to the touch of his fingers, and the feel of glass cloyed his palate like over-rich pastry ... Molten metal smelled like blows hammering his heart ...*[19]

Blind to its declarative crudity, I was beside myself with rapture when I read this at 14 or 15, especially since my own mental-imagery repertoire is almost non-existent. I can't even make a picture of a red triangle in my head, let alone one that tastes like a lobster. One genuine synaesthetic, Michael Watson, has been brain-blood-flow scanned and drugged and tested up hill and down dale. Watson specialised in linking odour/tastes with rudimentary but intense tactile impressions. The knack was beyond his control. Removing a chicken from the oven, Michael would be thrown into a tizz if it had too few 'points'—it tasted too 'round'.[20] Sucking spearmint under lab conditions, with amyl nitrite to enhance his sensory cross-link, and then amphetamines to dampen it, he literally felt the presence of smooth, glassy columns, projected outside his body.

Only one person in ten million is a full-blown synaesthete, according to neurologist Rick Cytowic, who studied Watson. Cytowic started by investigating synaesthesia and ended with a new model of the brain. High-tech tests suggest that we all make these translations from one sense impression to another, but it occurs at a computational neural 'level' prior to consciousness, and usually hidden from it. Cytowic tracked the activity to the limbic system, deep inside the brain. It only comes to consciousness when high cortical activity is, so to speak, switched off. Usually this happens only to people with disagreeable brain damage and consequent terrible deficits of awareness or ability. Rare, healthy synaesthetes like Michael Watson enable us to peek in and see the way our normally hidden or protected brain processes are partitioned and/or overlap.

Few topics are more enthralling, and infuriatingly evasive, than the brain and its workings. The human mind is supported by a brain that is parallel and multiplex in function, distributed rather than localised, with a reality-mapping cortex but a limbic zone which 'determines the salience of that information'. So a powerful emotionality lies at the heart of our humanness. It might well be that synaesthetes gain access, otherwise forbidden (except in 'out-of-body' hallucinations), to the processing of the 3-D 'model world'. This could be conducted in the hippocampus and other regions of the emotional and memory centres of the limbic system devoted to evaluating (or labelling) the salience, the relevance, of what we experience. Synaesthetes gain that forbidden access precisely by reducing the blood supply to the cortex, and flooding the deep core of the left hemisphere with rich energy supplies. Squirt radioactive tracers into the blood, and you can see it happen on the screen.

For Cytowic, this is proof that the self is primarily emotional rather than rational, located in the deep brain rather than the neocortex that covers it like crushed gift wrapping. Even more piquantly precise is Crick's suggestion that Free Will is localised in a portion of the inner brain, near the top and toward the front, called the anterior cingulate sulcus. Damage to this small group of cells is known to have caused an otherwise alert patient's mind to become 'empty', unable to communicate but unworried by that awful loss. It is a notion to cheer the shade of Descartes with his belief that the soul was attached to the brain at the pineal gland—and perhaps just as daft.

SPLIT BRAINS, UNIFIED BRAINS

The same general case is advanced by Michael Gazzaniga, the neuroscientist who taught us two decades ago about 'split-brain' patients. Gazzaniga argues forcefully for a modular account of both specialised brain circuits and, more explicitly, the sources of consciousness. He draws his theoretical inspiration from Jerry Fodor, philosopher of modularity. In the first place, he showed that right and left cerebral hemispheres have quite different talents. We blend at least two minds: one verbal and logical or interpretative, the other spatial and intuitive. This, of course, nicely mixes the ingredients commonly seen as male versus female. On a more detailed account, our brain's specialised modules (rather like the 'faculties' of an earlier philosophy) interact through an 'interpreter' region.

Most post-Chomskyan linguistics accepts this model for grammar; if Gazzaniga is right, it applies as well to 'the biological roots of thinking, emotion, sexuality, language and intelligence'. It looks very much, as Gazzaniga states, 'that the brain is indeed organised in a modular fashion with multiple subsystems active at all levels of the nervous system and each processing data outside the realm of conscious awareness ... These modular systems are fully capable of producing behaviour, mood changes, and cognitive activity'.[21] It's worth noting that other experts think this account is not the whole story. Willam Calvin and linguist Derek Bickerton declared: 'We have shown language to be innate, species-specific, supported by task-dedicated circuits, even if parts of those circuits may do double or treble duty in other tasks'—adding that 'brain imaging has shown that strictly locationist models of language function won't fly ... But then it was Fodorian psychology rather than linguistics that insisted on strictly localized, encapsulated modules.'[22]

Gazzaniga's continuing work asserts that our specialised brain architecture, which controls what we can know and do, is largely preset by the genetics that ordain each brain's wiring. He offers a Darwinian analogy of *selection* rather than *instruction* for the process of human learning. Australian Nobelist Sir Macfarlane Burnet's famous explanation of antibody formation showed how a thousand natural shocks prune or select a vast pre-existing array of immune cells. Just so, says Nobel-winning immunologist Gerald Edelman, each person's experience sculpts the wild forest of prenatal brain cells into the adult's neat garden of mind.[23] We are born neither 'blank slates' nor preprogrammed. Mind resembles the immune system rather more than it does a computer waiting to be loaded with data. From infancy we contain billions of variant antibody molecules, each primed for its antigen of choice. We don't *respond* to the world's impacts by devising cunning weapons of defence; we are limited to the armoury that we have carried from childhood. Adapting the principles of species evolution to the individual developing brain, Edelman's 'neural Darwinism' or 'topobiology' tried to explain consciousness in one mighty leap.

His audacity was breathtaking, for he had a strong opinion on every stage of the transition from brute matter to consciousness. From quantum theory to networks of nerves, from 're-entrant maps' that chunk groups of nerves, all the way up to language acquisition—oddly, he does not think highly of Chomsky's suggestion that we have an in-built 'grammar organ'—Edelman has constructed a testable, integrated model. Gazzaniga, however, admits that while 'the functioning modules do have some physical instantiation'[24]—that is, you can run a PET scan or a magnetic resonance, or cerebral blood flow techniques, and actually *pinpoint* the parts of the brain active during a given mental act—'the brain sciences are not yet able to specify the nature of the actual neural networks involved for most of them'. William Calvin's intriguing recent suggestion might answer that requirement, for arguably brains are 'Darwin machines' not just at the level of neural growth but in their operation as well. If Calvin is right, concepts fight for survival as quite literal feuding activation patterns dispersed across the brain's neural nets.

THE LANGUAGE OF THE MIND

Perhaps the most important result of these many and varied empirical and theoretical findings is that human consciousness, together with its predecessors and components, is in certain important ways localised rather than global, just as the cells and functions of the rest of the body chunk together into relatively autonomous organs. Even when the world comes at us in heavily preprocessed human language, we do not always find it easy to comprehend. Suppose you flip to the golden oldies station, and the Beatles sing 'the girl with colitis goes by'? Is this a weird acid flashback? Someone remarks, 'It's a doggy-dog world'. Then John Prine is singing, 'It's a happy enchilada, and you think you're gonna drown'. That can't be right! Ah: 'It's a half an inch of water', 'dog-eat-dog', 'kaleidoscope eyes'. These comic examples are from linguist Steven Pinker.

It is much worse when we try to understand some completely new aspect of the world, one not yet modelled by our customary fallible set of linguistic gadgets. Arguably, this is why science took so long to emerge, why it has done so only once in history (despite some honourable near-misses), and is easily shoved aside by inane but comforting superstitions. Even ordinary speech can be evasive, which is why we are baffled by this perfectly grammatical sentence: 'Buffalo buffalo Buffalo buffalo buffalo buffalo Buffalo buffalo.'

No, I haven't gone mad. A buffalo from Buffalo is a Buffalo buffalo, which might 'buffalo' or browbeat its kin. So '(Those) Buffalo buffalo (which other) Buffalo buffalo buffalo (will in turn) buffalo (still other) Buffalo buffalo.' This diabolical sentence is cited by Steven Pinker as an instance of the reach and limitations of our language 'organ', the DNA-specified mental 'instinct' that powers our speech and writing.[25] The buffalos' implications for the status of science as a special way of knowing are not self-evident, but they are worth teasing out.

Consider two bullets. One is simply dropped from near a gun's muzzle, the other fired horizontally, at the same instant, with terrific velocity. Which hits the ground first? Most of us have major trouble with this poser. Despite a lifetime of lifting and throwing and dancing, all our intuitions about motion start mumbling 'buffalo buffalo'. Most people conclude that the dropped bullet hits the deck first, probably much earlier.

Not so. Gravity pulls equally on both bullets. The sideways motion imparted by exploding gunpowder has no effect whatsoever on the rate at which the fired bullet falls to earth. How could it? But usually we need disciplined training in vector mathematics to understand this very elementary truth about how our world works. Science is *not* common sense. It is distinctly *uncommon* sense, and our brains—our minds—resist its enlightenment.

Recently, sociologists have asserted that 'western' science is just one form of many 'ethno-sciences', each with its own rich claim to be taken seriously as a form of valid knowledge. Alan Cromer is a fundamental particle theorist with a special interest in science education, and he won't have a bar of it. Most knowledge systems, he claims, project the culture-bound shape of human minds upon the outside world. In a special sense defined decades ago by the developmental psychologist Jean Piaget, they are 'egocentric'. By contrast, the techniques of inquiry invented by the Greeks and rediscovered three or four hundred years ago in Europe—techniques which have remade our world utterly—deny that outer reality can be known through intuition alone. While the daily practice of science is clearly swayed by rhetorical skills, special interests and power politics, it works so well because at base it strives for objectivity. In Piaget's terms, its practice requires 'formal operational' mental skills, which are never attained by more than half America's (and presumably all First World) adults. Hence, most of us 'can't analyze a situation with several variables,' as Cromer states scathingly, 'or understand a simple syllogism'.[26]

What of the intelligence of our evolutionary near-cousins, the apes? Steven Pinker has proposed Darwinian paths by which our mental modules evolved, including language itself. He is especially caustic about claims that bonobo chimpanzees have been taught a form of true sign speech. Pinker maintains that attempts to teach sign language or computerised lexical codes to chimps and other apes are doomed, because they did not share our eccentric evolutionary history. Human brains and other cultural organs, and the species-specific genes that dictate them, are precisely what make us volubly human and them merely clever apes.

But don't the apes share 98 per cent or even more of their DNA with us? True, but 'the recipe for the embryological souffle is so baroque that small genetic changes can have enormous effects.'[27] The contrary opinion, as we saw in an earlier chapter, is offered by the primate specialist Sue Savage-Rumbaugh, who argues that Kanzi, a young bonobo, routinely combines symbols using 'a primitive English word order [to convey] novel information'.[28] Pinker has tried to show that Kanzi is just aping language. Such disagreements, of course, are the very stuff of a science that strives for objectivity even when it studies the mind. Certainly the evidence put forward by primatologists has regularly been assailed as selective, wishful at best. Still, Savage-Rumbaugh and her husband Duane Rumbaugh have studied both common and bonobo chimps, notably Kanzi, for many years. In observations and experiments, they conclude that such primates can classify lexigrams (for example, into 'food' versus 'tool' groups), zestfully play computer games using screen and joystick (as can rhesus macaque monkeys) and, arguably, employ grammatical constructions.

TEMPERAMENT IN THE GENES

If the roots of language are inherited, as they certainly are in humans and may be in bonobos, what of temperament and other apparently less rule-bound aspects of personality? Although the Greek physician Galen got a great deal wrong, his theories of physiology and anatomy ruled western medicine for more than a thousand years. He is best known for his notion of the four humours—black and yellow bile, phlegm and blood—whose inherited dominance determined the patient's temperament: melancholic or choleric, phlegmatic or sanguine, or one of five mixed types.

The last living adherents of this quaintly old-fashioned doctrine were surely the Christian Brother teachers at my dismal technical school in the late 1950s, but it has been having an unexpected comeback. For decades, psychology taught that temperament, along with ability, was not innate but created by our experience of the world. Then a few years ago, with the discovery of neurotransmitters, we realised that brains and bodies are awash in chemical messengers, peptides, GABA, corticotropin releasing hormone and a hundred other tiny tides. The brain's logic circuits run as much on chemistry as on electric currents. And while our cortical and limbic circuits get tuned in an individual fashion, they tend to bunch into a few reliable categories, surprisingly close to Galen's humours. Recent research by Jerome Kagan, a Harvard psychology professor, has begun mapping temperament and detects strong heritability. His team's results, for all their nuance and circumspection, have inevitably stirred political passions.

There is good evidence that our feelings are organised in standard ways from earliest infancy, although styles of upbringing may skew the way we act out those feelings. Some 15 per cent of Caucasian babies studied were shy, timid or fearful, and these Kagan dubs *inhibited*. They tended to have quite specific physical characteristics: narrow faces, pale eyes, tall, thin, allergic bodies. Brown-eyed, smiling, fearless, robust *uninhibited* children made up 30 per cent. The rest fell somewhere between. (Chinese-American babies tended to be on average more inhibited.) Subtle new instruments and careful experiments endorse this pattern, and help explain the many acquired and hard-wired factors that build the profiles of temperament. Dominant activity in the left frontal brain relates to calm happiness, in the right with fear and sadness, while the right rear lobes govern emotional intensity, whether distressed or joyful.

Essentially, temperament seems to be a family of states organised around the level of reactivity of the sympathetic nervous system, which in turn is under the control of the amygdala, a part of the deep brain. High reactives are jumpy and reserved: worriers. Low reactives are extroverted, cheerful, fearless: warriors (or thugs).

THE CENTRALITY OF EMOTIONS

But are 'mere feelings' all that important? In the age of economic and other rationalisms, is it not rather soppy and Sensitive New Age to care about emotions? (Discourse analysts have noted that such objections are coded sneers at traits historically regarded as 'feminine' or 'queer'.) Actually, feelings are not only important to the quality of life but crucial to the human exercise of reason. Feelings, as Professor Antonio Damasio at the University of Iowa College of Medicine has made clear, are 'a window that opens directly onto a continuous updated image of the structure and state of our body'.[29]

René Descartes, you'll recall, supposed that we comprise a mechanical body yoked by divine contrivance to an impalpable and altogether finer self or soul, an immaterial essence designed to survive the corruption of the flesh. Since anyone could see that pure minds can't get angry or randy or choked up with sentiment, bodily feelings had to be downgraded. Now we know differently. 'The organism has reasons that reason must utilise,' declares Damasio, playing on Blaise Pascal's famous seventeenth-century phrase, *The heart has its reasons, which reason knows not of.* Thoughts or ideas are 'qualified' by feelings, which are markers within the body of how the world has affected us in the past. They are short cuts to *value*: powerful devices that help us nip through the waffle of unchecked logic. Victims of prefrontal leucotomy, whose links between the reasoning frontal lobes and the emotional amygdala have been cut, lapse into a feckless inability to plan or decide. They can know but not feel. And so their knowledge is short-changed, their reasoning not merely 'cold' but unhinged from reality.

Or is this analysis, in turn, a false polarity? Our culture's dearest tenet is the *irrational* source of the distinctively human, especially creativity and madness. Our world, according to John McCrone, is shaped by the legacies of Romanticism, with its split between wild, unchecked forces of nature (emotion, good) and wussy restriction (mind, bad). When Freud told us the irrational Id had to be throttled and harnessed by the workaday Ego and Super-ego, we mourned the repressive cost of unbuttoned joy. Artists, lunatics and Byronic lovers elude these strictures, and Lacanian psychoanalysis tracks our own deepest impulses, rather mysteriously, to endlessly deferred 'Desire for the Phallus' (which, unlike Freud's conjecture, has nothing much to do with the penis).

None of this is true, according to McCrone. We are made human by social language, as Russian linguist and philosopher Lev Vygotsky taught. Language completes our physical hardware (our bodies) with a specialised software (our minds). My voice, literally, makes me human— the voice(s) we each speak within our own heads, telling ourselves the story of our social world. Feral children, lacking exposure to language in the first crucial years, never became properly human. Deaf mutes denied non-vocal coding languages such as Sign suffer tragically impaired identity.

Madness and dream result not from unleashing some deep irrationality or the slithering signifiers of Lacanian psychoanalysis, but (by and large) from 'the broken-backed functioning of the bifold mind',[30] as McCrone puts it, the confused tussle of an inner voice trying to make sense of a faulty biological ground of sensation—disrupted neurotransmitters, in the comparable cases of madness or drugs. In ordinary dreaming, stray vivid memory fragments are juggled in sleep by linguistic machinery designed to narrate the tale of an organised external world.

THE POSTHUMAN MIND

Imagine, then, what dreams might plough the oceans of a brain scanned and replicated inside a machine, a mind uploaded into a computer in order to escape the strictures of mortality. Would it find itself mad or newly reasonable? Subhuman in its extraordinary condition, or trans-human, even posthuman? Having a computer substrate would surely alter the experience of memory, which now shows us its treasures through such a narrow window and holds them in so vulnerable a storehouse. Even without uploading, of course, the memory of the deathless might require substantial renovation and expansion, perhaps regular editing, or at least an improved means of filing and accessing its contents and links. Although our organic human brains do not process memory in any manner as crass as 'one neuron equals one memory' or even 'one con-stellation of neurons equals one specific recollection', there are surely limits to how much we can cram inside our ten or hundred billion brain cells and their webby synapses. One of the fruits of advanced cognitive science will be enhanced methods of storing and retrieving memories. For some of us, it won't come a day too soon.

Already, a blend of study skills and dedicated software agents can improve effective memory and even the power of our thinking. A software program called Remembrance Agent or RA, developed by Bradley J. Rhodes and Thad Starner at MIT's famous Media Lab, can run in the background of your computer and keep tabs on everything that happens on your screen. The program compares each word it sees with a growing cross-referenced database derived from your e-mail, the Web sites you visit, other selected files and data services. A small window at the bottom of the screen runs a constantly changing reminder of these links, and if you wish to consult one of the listed files you just click on it. When you close down your session's work, the RA updates its web of associations. As neuroscientist Anders Sandberg puts it: 'The RA combines the human, intuitive associative memory model with the efficient database memory of a computer.' Used inventively, the RA can act as a cognitive accelerator as well as an aide memoire to our grievously imperfect natural memory.

But will the first immortal generation wish to retain their (our) evolution-shaped and crippled limits? The prospects of a posthuman mind are both exhilarating and alarming. Will we even wish to conserve organic bodies? Consider this diverting anticipation by Russian-German Eugene Leitl (whose academic field is molecular modelling of wet, charged biopolymers such as peptides and lipids), a man not afraid to get right in your squeamish face:

> Apart from intrinsically higher and drastically cheaper diversity in artificial reality, a Dysonian computer cluster [where the planets of a star are reconstituted by an advanced science into a radiation-collecting shell with the surface area of a billion Earths] or a Jupiter brain is not a suitable environment for any classical body. Why limit oneself to human senses if I can utilize anything from proximal probes at the atomic scale to accelerometry, magnetometry, mass spectroscopy, particle detectors, huge interferometry arrays, etc? Exercise your own imagination on the motorics part. But, once again, if the material realm of a stellar system is entirely

under your control, maxing out on computation while keeping the number of resources allocated to sensomotorics at an absolute minimum (in a pinch, you could always re-use these atoms in new configurations albeit at a time penalty and resources wasted in dormant basal auto-replicator capability) is obviously a good strategy. Only coevolution-artefacted nonlinearities would seem to disturb the Brave New Postmaterial world scheme.[31]

Leitl (or 'gene', as he is known on the Internet) adds with further wolfish humour: 'I'd rather be a cluster of active-orbit-controlled boxes in high solar orbit swinging around my own gravitation center, while other parts of me are busily processing our stellar neighbourhood. (Of course there will be probably no 'me', just a memetic bouillabaisse virtually swirling in diverse bit-buckets of multiple shapes and sizes).'

QUANTUM MINDS

Yet even a brain the size of a planet or a Dyson sphere might not be enough to sustain a mind, if some critics of orthodox AI are correct. Recently, the British mathematician Sir Roger Penrose and his American colleague anaesthesiologist Stuart Hameroff have been exploring a quantum science of consciousness—or rather, trying to establish one. Mind cannot be understood and explained in crass computational terms, Penrose argues, without a new, extended kind of quantum physics that includes gravitation. This would incorporate the realm of relativity, Einstein's powerful mathematical framework of spacetime, which to date has not been integrated satisfactorily with the quantum realm of quarks, electrons and light. Penrose is a Platonist: he believes in a mathematical reality deeper and more primordial than the world we observe.

In several books dense with quantum theory and other tricky mathematics (allegedly simplified for the ordinary reader, but shockingly difficult nonetheless) he has tried to convince us that human brains can do things that computers never will.[32] If that's true, what is it about our mind/brains which makes us special? Do we, after all, have immaterial souls able to leap free of the restrictions that apply to matter? No, says Penrose, but we do have brains that might utilise mysterious quantum-realm abilities. His key is Hameroff's theory based on the allegedly strange properties of neural cytoskeletal microtubules, tiny hollow nano-scale tubes inside cells built from columns of tubuline dimers (protein polymers) able to switch between two conformations. Most physiologists see these microtubules as nothing more exciting than structural struts holding the cell nicely in shape. The quantum-mind theory speculates that their signal-transmitting properties might permit parts of the brain to act as cellular automata, or even non-locally (that is, contrary to the ordinary limits of space, time and causality) and thereby surpass their crass physical limitations.

It is a subtle argument, well beyond the scope of our discussion. Few specialists in any of the associated fields, however, are prepared to follow the speculations of Penrose and Hameroff (although they are attended to with a measure of respect). Interestingly, Penrose remains a reductionist, believing that one day we will possess theories robust enough to let us understand even the incalculable mind. For now, though, Penrose claims that minds *cannot* be emulated by conventional computers. If he is correct, mind will continue to be the proper, if baffling, study of a deathless humanity.

TAKING A QUANTUM LEAP

Armed with the conflicting findings of current neuroscience that we've just glanced at all too briefly, let us turn at once to the science which underpins Sir Roger's speculations. The first century of the new millennium might face not merely life extended without limit, which is the elaborate promise of genetic and nanotechnological engineering, but of life and mind utterly transformed by the mysteries of quantum theory. As you'll see, I have my doubts about some of the more popular claims in this realm. Yet it may well be the locale of major metaphysical upheaval—the place where science peels open not just the atom but the universe itself, which might turn out to be just one among a multitude or even an infinity of echoing mirrors.

five: quantum – many worlds & black holists

Granted that I have spent a considerable portion of my theoretical energies over the years criticizing reductionism ... it is worth spelling out again that to adopt a reductionist methodology in research strategy-that is, to try to stabilize the world that one is studying by manipulating one variable at a time, holding everything else as constant as possible-is generally the only way to do experiments from which one can draw clear conclusions.
Steven Rose, *The Making of Memory*[1]

Quantum this, quantum that, quantum everything.
Nobel immunologist Gerald Edelman scornfully dismissing 'surrogate spook' theories of mind (*Bright Air, Brilliant Fire*)[2]

Careless scuba divers are prey to a disorienting buzz: nitrogen narcosis. This addlement was known once by a beautiful, alarming name: 'raptures of the deep'. The compressed air mixture breathed by divers contains inert nitrogen, which gets forced into their brain tissues by the pressure of the watery depths, causing silliness, unmotivated giggles or, even more dangerously, fits of gloom and panic.

For most of us, untrained in the sciences, advanced physics is a realm no less alien than the deep ocean. Venture beyond the safe shores of the high-school lab and your common sense is put at risk. Like victims of nitrogen narcosis, you may find yourself slipping into irrational elation or despair. Call it, especially in its New Age or astrobabble forms, *raptures of the shallow*.

This is unhappy news if, like me, you take pleasure in today's cornucopia of scientific discovery. Science, after all, is the motor of our age, and a large part of its secret ideology. While it is distressing that so many of us are frightened, or dismissive, or simply ignorant of this ornate and fecund landscape, even worse is its gleeful colonisation by slipshod thinkers for whom equations are mantras. Those who find their latest shallow rush browsing the New Age 'holism' shelves especially adore books about cosmology and quantum theory—the majestically great, the enigmatically small—preferably if the words 'God' or 'miracle' are smuggled into the title.

For decades a key icon of shivery mystery was the black hole, that mysterioso realm found deep in interstellar space where the laws of physics get their comeuppance. Hijacked kicking and screaming from the tensor calculus-encrusted pages of impenetrable physics journals, black holes became sacred sites for 'holistic' ninnies blissfully gobsmacked by big ideas but too lazy to work at the hard details, or too eager for effortless epiphanies. Exploring the remarkable prospect of physical immortality puts us, unless we are very cautious, in danger of this unlovely fate, this black holism.

TOTEMIC ANIMALS

The most influential practical philosophies of the twentieth century comprised a piquant bestiary of heraldic animals, a symbolic performing circus. Ivan Petrovich Pavlov's wretched dog, pierced by tubes, ears pricked for the bell that would make it salivate involuntarily, stood in nicely for the former Soviet Union and other explicitly totalitarian regimes. B. F. Skinner's operantly conditioned rat ran a more liberal western maze while his superstitious pigeons, inanely repeating whatever chanced to coincide with a reward, foreshadowed the New Age. Sigmund Freud's Wolf-Man allegorised sexual confusion as slavering wolves perched in a tree, while Hans Eysenck argued that another Freud patient (little Hans by name!) had been terrorised into phobia by seeing a workhorse fall cruelly under its load.[3] Franz Kafka's beetle cowered in existential loathing before the shadow of Hitler's Reich, and Konrad Lorenz's greylag goose became trapped by ruthless genetic instinct into following any mother-figure it imprinted upon during its vulnerable infancy, and was cooked by napalm in Vietnam.

One way or another, these sorry beasts are constricting images of trapped minds, sullen and rebellious, at the end of their tether. How strange to reflect, by contrast, that the most profound mathematical upheaval of the century appeared to fling the cage door wide. Human beings, now neither insects nor dispassionate witnesses in white lab smocks, seemed to emerge from the science of quantum theory as frail, involved godlings, observers of a world they literally *created* (in some bizarre but fundamental fashion) in the act of observing it.

Or did quantum theory tell a contrary tale, that the core of reality was a lightless, roaring horror of uncertainty, lacking any determinate meaning?

SCHRÖDINGER'S CATCHPHRASE

Either way, the heraldic beast for this revolutionary philosophy was the oddest one yet, as wonderful as a unicorn and quite as fictitious. A ghost split between realities, a creature neither dead nor alive, it was known in the trade as Schrödinger's Cat. This luckless feline was the imaginary victim of a deeply troubling thought-experiment posed by the quantum pioneer Erwin Schrödinger.[4] (These days, when notices on car windows implore us to understand that 'Pets Are Forever' and 'Meat Is Murder', it has become customary to assure readers that such hypothetical experiments are conducted with pen and paper only. No cats were hurt in the making of this chapter.)

The hypothetical animal is confined in a sealed box—perfectly sealed from the rest of the world, which is a key to unlocking the eventual paradox—with a cyanide capsule which may or may not be smashed open by a poised hammer triggered by the decay of an unstable atom. The quantum physics of radioactive emission is governed by probability, rather than the kinds of direct cause-to-effect logic we are used to. So it can be arranged that there is, within the period covered by the experiment, on average just one chance in two that the isotope will decay and emit an electron. If it does, a detector registers the quantum event ... and the hammer falls. Instantly, the lethal gas suffuses the box and the cat perishes at once. If there is no decay, the cat is spared. What could be a simpler choice of outcomes, or a more sinister one? When we open the door, we'll find either a snoozing puss or a pitiable bundle of cold fur and bones.

Obviously. But what was happening to the cat *while the door remained sealed*?

Many people find their minds seizing up at this point in the analysis. Door? What's the door got to do with anything? Open or shut, the cat is either poisoned or not, either dead or alive.

Not so, argued Schrödinger, who thought he was performing a sarcastic *reductio ad absurdum* but instead kindled half a century of perplexity. Poor pussy's fate is intimately linked with the state of the radioactive isotope inside the sealed box, which—quantum theory assures us—remains *both* intact *and* decayed, in a condition of probabilistic suspension, so long as no measurement is made of the state of the box from outside its confines.

We are not discussing scholastic theology here, or Jain multi-valued logic. Keep this truth at the front of your mind: the formal mathematics of quantum theory *work* in the most practical way. They describe and explain and predict states of the real world (the electronics of computers, in particular), with finer accuracy than any other system of knowledge humans have ever devised.

That QT formalism insists, against all orthodox reason and common sense, that the radioactive atom inside the box subsists as a mixture of overlapping or superposed waves. (More properly, something called wave-function, but we'll let that slide.) One wave stipulates an intact atom—and that tiny, microscopic state of affairs becomes magnified up to the macroscopic realm, where we live, *as the palpably alert cat.* The other wave specifies a decayed atom, and in this case, because the atom and the rest of the prepared box are muddled together, its microscopic state is amplified, sadly, into a dead cat. The wave-functions of both possible outcomes are *entangled*. Neither is sufficient, alone, to describe the condition of blended box, isotope, hammer, flask of cyanide, cat ... until the instant when the system is, as they say, 'observed'.

OBSERVED?

At that moment (but which moment exactly? observed by whom, exactly?), the superposed waves are said to *collapse* into a single, definitive description. The whole awful box and dice stops being probabilistic and snaps into certainty, with probability one point zero. Either: *dead cat*, and keep your gas mask on, doctor, there's still cyanide leaking from the opened box. Or: *live cat*, here puss, puss, have some milk in your favourite bowl. (Or, in some staggeringly unlikely case, no cat at all, because the animal has 'quantum tunnelled' out of the box and run off as fast as its legs would take it. But we'll ignore that possibility as well.)

Here's the key, once more: the infinitesimal effects at the level of a single atom, its protons and neutrons and orbiting electrons, are correlated with the health of the cat and, since both states of the cat are entangled and overlap, the animal has been *neither* dead *nor* alive.

Until observed.

Well, that is, as you have probably noticed already: unless the cat observes itself. But sadly, no, this is just a trick of words and not a solution to the paradox. What's needed is an interaction from *outside* the box.

Or unless the mutual impact of all the whirling particles that comprise the box and dice observe each other, so to speak, by banging into each another, bouncing hither and yon, leaking away from the surface of the box into the external world, absorbing heat and light and sound from the laboratory ...

Or does the observer need to be altogether outside the system? And what observer would that be? God?

Or—and here's the rub—a different *kind* of thing from a cat or a box or an atom: a mind, let's say. An *observing mind*. A *consciousness* that, in some unexplained way, is magically free of the constraints of matter and energy, able to 'observe' in a locally godlike fashion. How New Age! This extreme possibility, followed rigorously, led physicist Eugene Wigner to conclude that human consciousness is the crucial determinant of reality. Wigner, by the way, was not drummed out of the halls of science. He was given a Nobel Prize in 1963 for contributions to nuclear physics (although not for this particular idea).

For Wigner, the universe did not really exist prior to the emergence of intelligent life. It is a theme that was endorsed and elaborated by another immensely influential quantum theorist, John Archibald Wheeler (the man who put the term 'black hole' into circulation, although he did not coin it).[5] The risk, as I've noted, is that these dazzling studies are all too easily conscripted by quantum quackpots and black holists. A chapter in a recent book by the theorist Murray Gell-Mann is entitled 'Quantum Mechanics and Flapdoodle'. He cites with relish Stephen Hawking's dictum: 'When I hear about Schrödinger's Cat, I reach for my gun.'[6]

MIND AS A CAT'S CRADLE

But might such a preposterous theory indeed explain the mind itself, which the previous chapter asserted can be explained purely in materialistic terms? One impressive attempt to unite consciousness and rarefied theory has been offered by David Hodgson, not a physicist but an Australian Supreme Court Judge.[7] Trained as a philosopher, Justice Hodgson taught himself quantum theory on the long train trip home from work, and argues the case that mind and even deity lurk in the Schrödinger equation. His exploration of paradox and puzzle in the quantum realm resembles a cool judicial summing-up. He punishes the 'many worlds' version of cosmological quantum theory in which the universe is supposed to 'split' whenever alternative or superposed quantum pathways branch—but points also to discrepancies in the evidence for other interpretations. I find it impossible to agree, however, with his religious verdict that mind transcends and 'collapses' quantum probability-states into the reality of matter, that 'truth may be allegorized, and thus in a sense approximated, in different ways, which are mutually inconsistent, but which are all approximations to the unexpressable truth'.[8]

But is my revulsion for this ruling no more than a prejudice in favour of traditional hard-edged logic and rationality? It's true that quantum theory seems to prove our world is just an averaging of radically indeterminate parts. Gell-Mann and his colleagues believe they've found a more rational way to interpret the quantum equations, invoking what's called 'decoherence'. This difficult topic is at the heart of Gell-Mann's current Sante Fe reflections on simplicity and complexity. But it is typical of Gell-Mann's rather pushy omniscience that he knows Hawking's jest about reaching for his gun does not derive from Goebbels, as most of us suppose. No, the chilling gag about *Kultur* and Brownings came from 'the early pro-Nazi play *Schlageter* by Hanns Johst'.[9] So much for the notorious gap between the two cultures. By a sort of inevitable irony, quantum indeterminacy, as a crippled metaphor, has been imported into the humanities as deconstruction, another kind of theory disputing traditional boundaries and verities.

Murray Gell-Mann is a Nobel laureate who devised and named today's standard quark model of elementary particle, but also, as noted, probes complex adaptive systems such as rainforests, economies, and us. He is one of a number of physicists whose names are becoming known to the wider world, people such as Stephen Hawking, Steven Weinberg, Kip Thorne, Paul Davies and Michio Kaku, all of them mathematical physicists whose research blends the tiny world of quark and electron with the far end of the universal scale: massive black holes where space and time are extinguished. Others, who tend to be philosophers well versed in arduous physics, take the final plunge into metaphysics, debating whether these epochal upheavals tell us something new about the purpose, if any, of creation.

Despite the harrumphing, there's nothing meagre or killjoy in Gell-Mann's universe. He's as much concerned with the musky, muscular reality of a jaguar in the forest as with the rarefied quark. And his theories are gratifyingly wild. Beneath the carefully defined jargon, strange possibilities lurk. Quantum physics, for Gell-Mann, is a way to track divergent, alternative histories of the universe. Do those domains truly exist, so to speak, as parallel universes? 'Could an observer utilizing one domain really become aware that other domains, with their own sets of branching histories and their own observers, were available as alternative descriptions of the possible histories of the universe?'[10] While Gell-Mann regards it as an open question, supersymmetric string theorist Michio Kaku is bolder.[11] Unified physics deploys many dimensions beyond the four of Einstein's spacetime. The best available model for the grammar of reality uses 'heterotic strings', vanishingly small loops that vibrate clockwise in ten dimensions and counterclockwise in 26 (16 becoming 'compactified'), or perhaps extended manifolds or membranes. Is this an improvement on staid old Newtonian absolute space and time? Yes, as beneath its surface complexity it permits wondrous mathematical elegance, simplifying the many laws of nature into a single geometry of fields in hyperspace.

The superstring explanation replaced quarks, leptons and their force particles (which had replaced mid-century accounts of the atom as those had replaced the aether) first with those inconceivably tiny wiggling loops whose vibrational rates specify which particle they are at the moment, and then with a kind of 'membrane' that subsumed them in turn. While superstring theory represents a striking simplification of physics (if accepted), it does so by giving the universe nine dimensions of space and one of time. We perceive only four dimensions because the other six are 'rolled up' out of sight. Still, because these vibrating primordial objects, these smallest of all things, now have a kind of length and perhaps breadth as well—they're not pure geometrical points, as used to be supposed—certain noxious infinite quantities cancel usefully out of the equations. Many mathematical problems that dogged the search for a general theory of all spacetime and its contents are solved in a trice.

STRINGS AND MEMBRANES

Invented in the late 1960s to explain the strong nuclear force, the original string theory was temporarily abandoned when Gell-Mann's quarks did the job better. A decade later superstrings had a comeback, shrunken in size to one hundred billion billionths of the scale of a nucleus. I am no mathematician, so I have followed this vivid emerging tale in a remote, admiring and allegorical way by scouring the words of its founders: people such as John Schwarz, Edward Witten, Nobel laureates Abdus Salam, Sheldon Glashow, Richard Feynman and Steven Weinberg. Some of them hailed superstrings as the key to the new physics. Others deplored it as a fad. Currently it remains the best candidate for Theory of Everything, somewhat transformed into 'membrane' super-theory, known for short as M-Theory.

String theory mathematics, it has been claimed by the discipline's chief inventor Ed Witten, is twenty-first century physics that fell accidentally into the twentieth. Even today's finest minds are not quite up to the job of using this powerful new tool definitively. Oddly, current research opens the possibility that information might *literally* fall from the future into the past. Both Kaku, professor of theoretical physics at City College of the City University of New York, and Kip S. Thorne, Feynman professor of theoretical physics at Caltech,[12] have described wormholes in spacetime, solutions in quantum gravity theory that promise to link distant regions and epochs in the universe.

Whether such wormholes are physically plausible, and not just annoying quirks of the equations, is an unanswered question. Until very recently, wormholes and time machines ('closed timelike curves', in physics jargon) were ignored by nice people. Now it look as if they might well be the basis of technologies available to our descendants— for stellar engineering feats, using 'exotic matter'—or to other advanced civilisations elsewhere in the cosmos. The prospect is enthralling, although it might unhinge the chronically prosaic.

The end of this road, flagged cheekily and profitably by Stephen Hawking in his famous quip about 'knowing the Mind of God', is the question of the origin and destination of the whole universe. Even if our local universe is just one within a vast manifold or sheaf of superposed alternatives, or temporal loops, it had to start somehow. Is it reasonable—even necessary—to posit a deity who kickstarted reality? If so, must this be the God of the Bible (or the Koran, the Moonies, or perhaps David Koresh)?

While superstrings get all the headlines, an alternative Theory of Everything has been beavering away quietly in the background. Invented by Hawking's collaborator Roger Penrose, *twistors* also dwell within more dimensions than we're used to, though Penrose rations himself to eight. I must repeat my confession: unless you're a mathematician, this kind of information tends to skid straight down your brain and off the tip of your nose. But the experience of dabbling in advanced physics is fun, and although at the end you are likely to be as ignorant as you were at the start, it's a far more *exalted* kind of ignorance.

After such extravagant vistas, can science continue to find fresh ways to astonish us? Of course. Scientists have recently observed quantum entanglement, for example, using entire atoms rather than individual particles or photons. Highly excited rubidium atoms have been quantally entangled with atoms in the normal state, and it has been shown that measuring the atom in the ground state does, as theory predicts, instantaneously alter the exited atom, or vice versa—even though their connection is *non-local*, that is, the defining interaction has occurred faster than light. It is expected that these effects will soon be duplicated in entire molecules.

The *next* and even more unimaginable physics presses forward into realms traditionally regarded as science fiction, such as time travel and 'parallel universes'. Prestigious theorists have lately provided grounds for taking these topics seriously. Such speculations spring from the awful realm of the black hole, where near-infinite gravity smashes time, space and regular physical laws. Here time can be reversed, and with cunning navigation perhaps one might live to tell the tale. Thorne and his associates have shown that in principle 'quantum wormholes' can serve as gates through time and space. These are exactly the kinds of discontinuities one might expect to emerge from the accelerating science and technology of the next centuries, and make it all the more enticing that we might live to see it.

Careful logical studies show that self-consistency is the key to plausible time travel (either physical, or via messages to the past). In effect, the blurriness and uncertainties of the atomic world become amplified up to our own scale of existence, saving the time traveller from paradox—but perhaps only at the cost of swapping one history for another! Don't expect to see a time gate opening at a store near you soon—they require fabulous engineering and 'exotic matter'. Still, one might wonder if information slipping through wormholes explains seemingly impossible results from parapsychology and brain-scan labs. According to Dean Radin, scanning records in at least one lab have shown that people make a kind of prophetic response—a *presponse*—to powerful stimuli that they will not actually see for another second or so.[13] That is, physical reactions in the brain have been recorded (or so it is claimed) in which subjects unconsciously reacted *before* they were shown emotionally charged images. It seems plausible that such effects are due to quantum weirdness.

DOWN TO BASICS

The rumour that science is on the verge of a Theory of Everything has attracted more than a measure of scorn, and can ruffle the easily upset. Whose bailiwick is safe? Watch the cutlery! Fears about the inroads of a Final Theory have a political dimension, and have been addressed by Nobel laureate Steven Weinberg. He won his 1979 prize by unifying two of the four fundamental and apparently distinct forces of nature. Now, he believes, we are in sight of a solution encompassing all four. Perhaps everything is built of tiny rips or 'glitches in spacetime'. Indeed, 'in the latest version of string theories space and time arise as derived quantities'.[14] A final theory would require its mathematical model of reality to be so savagely restrictive that not one variable can be altered without crippling the whole glorious construction.

'Once nature seemed inexplicable without a nymph in every brook and a dryad in every tree,' Weinberg has remarked. 'There are still countless things in nature that we cannot explain, but we think we know the principles that govern the way they work.'[15] So we are in no danger that a final theory of physics will try hubristically to claw back the domains of biology, psychology or art. Weinberg is not, as evolutionist Ernst Mayr called him, an uncompromising reductionist. Wryly, he dubs himself a 'compromising reductionist'.[16] What this means is a hope and faith that every question 'Why?', every separate arrow of explanation, points to a common denominator: the strange but rigorous laws of quantum mechanics. And yes, 'the reductionist worldview *is* chilling and impersonal. It has to be accepted as it is, not because we like it, but because that is the way the world works.' Nor is this explanatory realm of science alien to our everyday experience. Why is chalk white? Weinberg has shown, in a series of steps, that at every level of explanation of such a simple phenomenon as the whiteness of common chalk we are pushed back to quantum theory, and perhaps beyond that to a final theory. (His subtle account of why this is the case updated Thomas Huxley's famous lecture a century back to workingmen in the chalky region of Norwich.)

'Physics is the most pretentious of the sciences,' Paul Davies has remarked disarmingly, 'for it purports to address all of physical reality.' From snowflake to galactic cluster, all is controlled by bureaucratic regulation. 'The physicist believes that the laws of physics, plus a knowledge of the relevant boundary conditions and constraints, are sufficient to explain, in principle, every phenomenon in the universe.'[17] But science, despite these almost theological ambit claims, is by no means written in stone. Quite the reverse. Physicist Chris Isham has observed that while it is very exciting to be involved in creating a theory that works, 'in many respects it is even more exciting to be present at the collapse of one of the great edifices ...'[18] We confront just such a collapse, as relativity and quantum theory face each other off across intergalactic gulfs and the windhover blurs of electrons. The news is not necessarily good.

'Contrary to conventional wisdom,' Harvard astrophysicist David Layzer has declared, 'there is a vast amount of genuine, irreducible randomness in the world.' But quantum processes, inherently indeterminate, are *not* thereby more 'creative'. 'The outcomes of self-creation are new and unforeseeable, yet coherent with what has gone before.'[19] The contrary of creativity has always been noisy disorder, and the worst form it takes is death itself, the final spectre at every feast, including science's overladen tables.

DEATH IN A QUANTUM UNIVERSE

Life's a bitch, snarls the T-shirt slogan, *and then you die*. Bitterly trenchant, it seems to capture the worst possible take on our fate: suffering, and aware of it, in a heedless cosmos. Even less optimistic pictures are available, of course. A Hindu cynic might complain that *Life's a bitch, and then they send you back to go through it again.* And during the longest stretch of our common Christian heritage, the gloomy knew in their bones that *Life's a bitch—and then you go to Hell and burn forever.* So death as obliteration is not the worst fate we phobic, self-punishing creatures can imagine. True, hell or rebirth into misery are the downside options of the remedy we invent for the pain of transience, the loss of loved ones, the certainty of decline and extinction. It is hard to know which came first: our wistful fabrication of an eternal afterlife where the injustices of an imperfect world finally would be set right, or our trembling terror at unremitting torment for the unpunished ills of our mortal span.

So we deal with our unprecedented knowledge of death's inevitability (to date, at least) by contortions of denial and ingenuity. Actually, most of us conclude, we don't *really* die. No, we're recycled, or go on to a better place, or transcend rude matter by a dozen baroque routes. Today, these creative adjustments to the books are about to be audited. The universe itself, as we shall see in greater detail in the next chapter, turns out not to be eternal, but has a beginning at the Big Bang, a middle governed by inexorable entropy, and an end either at the Big Crunch, when everything is swallowed by a whopping black hole that extinguishes all the properties of time and space which make for individuality, or in the exhaustion and dissipation of the Big Whimper. Perhaps we need a new theodicy—a fresh justification of the ways of eternity to humankind.

One version increasingly encountered, depressingly, is the pseudo-quantum song warbled by smiling New Age irrationalists. Quantum theory lends itself to such appropriation, being grounded in that Schrö-dinger link between observer and observed which classic physics overlooked. Quantum observership can be hijacked into a plea for the special pre-eminence of consciousness in 'creating reality', usually mixed with the striking truth that 'reality' is in part a social and linguistic construct, ceaselessly revised by political negotiation.

Veteran physicist John Archibald Wheeler, who is usually brought on stage in support of this misunderstanding, in fact spent many years denouncing his traducers. A quantum observation, as we have noted, is most rationally seen as any definitive interaction with an ensemble of atoms which produces decoherence. When a single particle mysteriously passes through two slits and interferes with its own possible pathways to create wave patterns on a film, it is the *film* that performs the observation, *not* the lab technician who develops it. Unless, of course, you prefer the quantum 'Many Worlds' hypothesis. You'll recall that this doctrine, a favourite of Stephen Hawking's, holds that the famous principle of quantum randomness does not reflect an inability to decide atomic facts, but registers a 'splitting' of the universe each time a micro-level choice is determined.

Of course, this finding from the micro-world does not apply at the large scale of free human choice, though fashionable mystics coopt it to that end, failing to see that this destroys moral accountability. Regard the logic: your decision not to rape, steal or pillage must be matched (in another split-off universe) by its alternative reality where 'you-2' have chosen the crime. Why miss out on all the ugly fun? If *someone* is bound to tear up the bar, or embezzle the life savings of a sobbing widow, it might as well be you. Leave dull virtue to your quantum double. I don't think so.

THE MORALITY OF MANY WORLDS

It might seem a curious step to fault the Many Worlds quantum model for its dire *ethical* implications. After all, we don't reject the validity of nuclear physics because it allows the detonation of thermonuclear weapons. On the other hand, a perfectly good motive for rejecting psychological behaviourism (now largely defunct, I'm glad to say) was its denial of the experiences of consciousness and moral choice. When I act well or ill, it is not because my superspace multiple-doubles choose otherwise.

Worst of all, black holists tend to confuse the Problem of Evil with the Problem of Pain, so that their proposed *karmic* solution to the former resembles the ingenuous gym sign: *No Pain, No Gain.* I suspect the victims of any number of death camps would feel less than consoled if assured that their brutalisation was designed (or even permitted) in order to kick the rest of us up some developmental ladder to a renovated and 'more highly evolved' estate.

What, then, of death itself, the general and universal spectre (until immortality puts it on hold) at the feast of life? Sir Fred Hoyle has suggested that time-loops allow future actions and states to influence the past events which, to our time-trapped egos, gave rise to them. Now this, oddly enough, is by no means as absurd as it sounds.[20] As noted, recent theoretical work by Kip Thorne and associates show how reversed-time trajectories are possible without toxic consequences. Wheeler himself, astonishingly, proposed that 'the universe, through some mysterious coupling of future with past, require[s] the future observer to empower past genesis'.[21] As I've noted, well-designed pre-cognitive parapsychological experiments at Princeton, Edinburgh and elsewhere also seem to require such time-bending effects (to Wheeler's outrage, ironically enough).

This is a far cry from the animism of, say, the rogue quantum theorist Fred Alan Wolf. Here is one Wolfian gem I especially relish: 'The photons in [a] laser tube are bosons and so tend to "psychically condense" into the same state. This "boson condensation" is the physical manifestation of a universal and very human quality—the feeling of love.' So laser light is driven by ... *lerv*! Doesn't it just make you go all warm and squishy? But, hey, this is profound, man. 'The human body-mind is that autonomically functioning aspect of spirit, or "qwiffness" ['qwif' is Wolf's coy acronym for 'quantum wave function'], which is ultimately the form and body of God.'[22] Oh, right.

WOOLLY MASTERS DANCE

Black holism is mushy theft of hard, rigorous, limited, testable ideas from science. Granted, science is based on metaphor, of course, like all discourse, but that fact should not license the warping of its hard-won findings by those who wish to shore up their angst with a miasma of quantobabble. Discussing John Bell's non-locality theorem, the late quantum theorist Heinz Pagels noted scornfully: 'Some recent popularizers ... have gone on to claim that ... the mystical notion that all parts of the universe are instantaneously interconnected is vindicated ... That is rubbish.'[23] You'd never know it from the happy zeal with which black holists wave the Bell theorem like a mantra. Of the ensemble theory of quantum reality, Pagels asks: 'And didn't John Wheeler, one of the physicists who helped develop this many-worlds view, finally reject it because, in his words, "It required too much metaphysical baggage to carry around"?'[24] Introducing a magisterial biography of quantum pioneer Niels Bohr, Abraham Pais wrote recently: 'I hope that the present account will serve to counteract the many cheap attempts at popularizing this subject, such as efforts by woolly masters at linking quantum physics to mysticism.'[25] (*Wu Li Masters*, indeed!)[26] Anyone with a grain of sense and taste will know what to do with such claims. Certainly they won't erect a theory of Life, the Universe and Everything on such a corroded armature. The salience and gravity of these themes—cosmos, consciousness, death—is what makes such muddled syncretism so deplorable.

Recall that physicist and educator Alan Cromer has argued that science can really only be done by people with formal operational skills (mental abilities defined by psychologist Jean Piaget) which inadequate schooling has failed to develop in perhaps half the adult population or more. Without these tricks of objective thought, Cromer and Piaget argue, humans construct egocentric narratives about the universe based on projecting the inner world into the outer. Ego-centric in the literal sense—their worlds are experienced as if actually projected outward from their own minds, as if *directly* malleable before the power of wish. (Of course not even the hardest-nosed reductionist doubts that optimistic or self-paralysing attitudes can skew the way you *act*, which in turn changes those parts of objective reality you deal with and modifies the attitudes of other people.) Egocentrism is the assumption about the supposed top-down flow between (a, your) mind and the world, which is made explicit in the sad little New Age slogan: *You create your own reality*.

Another typical move in black holist rhetoric is to speak of 'highly evolved human beings'. More than a century and a half after Darwin, this is an amazing gaffe. It is by no means self-evident that one may sensibly speak of one *species* being more 'highly evolved' than another. (By what standards? If an HIV virus kills a human after replicating itself successfully, does its success prove that it is higher up the food chain and hence 'more evolved'?) Stephen Jay Gould has made a minor career of turning this lazy, anthropomorphic conceit into mincemeat, most recently in an extended argument showing that for three billion years bacteria have been the very model of a modal major life form. But still, even if we allow that by some measures of complexity one species may meaningfully be elevated over another, it is populations of *genes* that evolve, never *individuals*. Imagining that individuals 'evolve' in their own lifetime is an ignorant holdover from Theosophy at the pre-Mendelian turn of century (and even then people were supposed to 'evolve' via a series of reincarnations, which *almost* makes sense, although not Darwinian sense).

Nor is this misuse of scientific language a mere lapse among those who wish to find a more portentous word for 'grow' or 'mature'. The implication is plainly that 'evolution' is a goal-directed activity (as maturation certainly is), and that it is drawing us toward a predetermined transcendence. That implication is as wrong as one can get.

The non-intentional, self-organising properties of the universe emerge from the 'bottom up', out of the properties of its small stupid parts, modified by the presence in the local neighbourhood of lots and lots of other tiny mindless parts. This is true even if Stuart Kauffman and his Santa Fe colleagues are correct and complexity tends to emerge via a quite limited and fairly inevitable number of pathways. The interactions of many small, stupid things tumbled together may form 'attractors' in an abstract configuration space, but those mathematical attractors do not *cause* the behaviour they map and predict—not even to the extent that roads restrict and channel the routes we drive along.

Beyond Holism and Reductionism

Things in the world accrete by the principle of the ratcheting crane, to recall Daniel Dennett's neat metaphor, and not by being tugged upward into the empyrean by some holistic skyhook. Once their parts are stuck together they often form new 'chunks' with a surprising and novel integrity, modules that thereafter manifest new and determinate properties not visible in their elementary parts. But this is because structure *limits* the degrees of freedom of those constituents.

My table holds up my computer because its small, stupid atoms lock together in a fairly stable, robust fashion, which they do by restricting each other in their mutual proximity. Is that a holistic fact or a reductionist one? Neither—it is an explanation of a local whole by an understanding (in broad, general terms) of its nanomolecular bits and pieces. I am not talking about the human intentions which surely were crucial in selecting, shaping and connecting the pieces of wood into a table. Nor do I mean the informational unfolding of the tree's DNA which turned organic chemicals into timber. These organised patterns, wonderful as they are, remain parasitical upon the laws of quantum physics that obtain at the foundations of phenomenal reality. And our knowledge of those laws, and their physical manifestation, are utterly reductionist.

Nobel laureate Steven Weinberg cites Dostoyevsky's underground man: 'Good God, what do I care about the laws of nature and arithmetic if for one reason or another I don't like these laws ...' Weinberg's scorn for this self-satisfied despair leads him into an awful pun: 'At its nuttiest extreme are those with holistics in their heads, those whose reaction to reductionism takes the form of a belief in psychic energies, life forces that cannot be described in terms of the ordinary laws of inanimate nature.' He adds, candidly: 'The reductionist world-view *is* chilling and impersonal. It has to be accepted as it is, not because we like it, but because that is the way the world works.'[27]

Why, then, do so many people so ardently believe otherwise?

Our supreme eagerness to impute purpose, telos, *narrative* to the world arises because that *is* indeed an appropriate way to understanding the *human* realm, the regime of intentions for which our minds have evolved their special competence. Minds, as we've seen, are more like rowdy parliaments than lawful diamond crystals. At the same time, like parliaments they operate through a selected speaker. Oliver Sacks' *An Anthropologist on Mars*, for example, is rich with bizarre and illuminating tales about the mind and its breakdowns, especially autism and autistic gifts. He gives powerful evidence for mental modularity, while insisting that what is missing in autism, say, is exactly a higher level of global organisation—a categorising *self*. Is this holism or reductionism? The complex reality, in a way, eludes simple labelling. Finally, though, we are faced with wounded people whose global functions have failed to emerge, one way or another, from the parts that comprise them. Yet clearly we are genetically structured to work as 'selves', despite recent deconstructive bids to disperse the self.

At a recent seminar given by a young woman graduate student working on cyborgs, cyberspace and feminism, I heard much citation from Luce Irigaray and other poststructural gurus. The performance was dazzling, impressively learned after its fashion, and … totally arbitrary. No method was invoked for testing the assertions other than by their shock value, on the one hand, and their adherence to fashionable doctrine, on the other. When I asked why possibility X or Y had been discarded, I was told that the speaker just didn't think it was … didn't feel it was … didn't … Next question, please. I pressed her. Roboticist Hans Moravec claims X, I pointed out, and he's one of the world's specialists in this field of artificial intelligence, so how can you just bypass his claims because you don't *feel like liking* them? Er um, what I really wish to concentrate on is …

The road to black holism is lined with cognitive dissonance.

MEDICAL NEMESIS
Another example: I was once introduced to a New Age guru who, an acquaintance assured me, was charming and spiritually wise. I found him likeable, as advertised, sickly sentimental and utterly gullible. We had a very long and painful conversation, interspersed by frantic cups of tea. Each statement he made was weirder and more credulous than the item preceding it; the world was run on principles of numerology, and you can change traffic lights to green by wishing hard as you approach them, and aromas cured everything that ailed your body (except for the grave intestinal illness that had almost killed him a few months earlier, fixed in hospital at the last moment by huge doses of cortisone—which he ungratefully and in retrospect dubbed 'toxic chemicals' forced upon him by a crudely reductionist science), and the iris was connected to da *hip*-bone, and UFOs were driven by little grey aliens, and Sai Baba could levitate and pour endless quantities of dust out of his open hand and

vomit up gold phalluses a foot long, and crystals had magical powers, and the Urantia Book was the source of all wisdom, such as the passages about how humans had evolved through the stages of Green Men and Purple Men and Blue Men, and ...

At first I listened to this gibberish with wonderment, as if privileged in an anthropological way to observe a rare subculture of Pre-rational Humankind. After a while (shame on me!) I gratuitously began to lecture the poor man, in an increasingly frenetic and cross-disciplinary way, on alternative explanations for these odd phenomena, the sort that current science might put forward. I ranted on, delivering myself of cognitive science explanations for just *why* he found such unsubstantiated hogwash so believable. He reeled away hours later in a bruised mental condition, I gather, and I have never seen him since.

You might wonder, as I do: why did I feel impelled to *set him straight*? Leave aside whether I have any warrant for thinking that my book-larnin' gave me a grip on truth superior to his woolly word-of-mouth subcultural melange; of course I do. But I don't stand on street corners preaching godlessness to passing Mormons. This urge only comes over me when I find myself in a room with one or more black holists whose views affront me in their inanity and intellectual poverty. Or, more importantly—not just their views, but their ways of deploying, testing, applying and revising those views.

Despite appearances, I am not especially dogmatic about any particular subsections of my world view. Like many intellectuals struggling to keep their heads above the torrent of new books and journal articles and Internet postings, I modify my opinions about quantum theory and cosmology and the structure of mind and society according to whichever brilliantly argued source I read last. But (no credit to *me* in this fact, of course) those somewhat flexible views form a kind of mutually bracing geodesic structure of some power and authority. So maybe they generate Dawkins-style memes that insist on broadcasting themselves and fighting opposing memes to the death.

Is holism a meme deserving of death, or at least retirement? On the one hand, as we have seen, in some misleading sense *everyone* is a holist. Who would deny that the simplest protein or even molecule manifests spontaneous self-organisation, folding, specificity, that could not easily be predicted from its components? That is even more strongly the case for living creatures and cultures. But none of it implies there's something like a *mind* doing the universal shaping, putting in the complexity from the top down. And that is the crucial explanatory step for most holists.

Nor is emergent mentality at a level higher than the individual human especially consistent with the world we observe. Despite alleged 'synchronistic' events or coincidences, you recall *your* personal experiences rather than *mine*. We simply don't remember the experiences of others, although we do routinely fabricate interior models of those people who are important to us. We mentally 'act out' their probable experiences as a kind of inner drama. The bottom line is this: when I feel an itch, and raise my hand to scratch my nose, your arm does not lift into the air instead. We are firewalled by our skulls. If there is a 'holistic' channel, it is clearly very weak and noisy.

Top-down 'holistic' lines of analysis reek to me of sad reports of African Ebola victims. Owing to their ancient holistic wisdom, the luckless victims in Zaire *know* that the disease is caused by evil or enraged spirits, having nothing to do with anything so stupidly reductive as an infesting micro-organism, so they're holistically bribing the guards who so unreasonably lock them away from the free exercise of their right to go hither and yon unchecked, spreading the damnable disease.

Granted, that isn't a very *sophisticated* brand of holism, and classy black holists can cite Paul Davies and Robertson Davies and for all I know Bette Davis in their cause—but I believe it remains wrong-headed for just the same reasons that disease-demons are. The nearest I have seen to a strong scientific case for physical holism is John Gribbin's recent book *Schrödinger's Kittens*, which promotes John Cramer's quantum theory as the Final Answer.

Cramer, a physicist at the University of Washington, proposed a 'transactional analysis' or 'absorber theory' of how the wave function collapses when a quantum state is measured. He sees this as a kind of handshaking across time and space that gives a literal meaning to phenomena usually treated as mathematical fictions. When an event occurs at the microscale of the quantum, mathematics describes it as a blend of 'retarded waves' (which spread out in time and space, like a ripple in a pond), and advanced waves with negative energy that travel backwards in time. Usually physicists ignore the advanced waves. Cramer, though, builds his unorthodox theory on them.

For him, the retarded part is an 'offer wave', flowing out into the universe in the customary fashion. When it is absorbed by running into some atom, an advanced or 'echo wave' is fired back along the same path, a pulse literally flowing backward in time to meet the offer wave and interact with it. Via that encounter are created those real, observable quantities we find in the laboratory and which underpin the existence of everything else.

Much is made of non-locality, of the strange connections from present to future and back to past that wind everything together in the Cramer model. Sceptics are obliged, learning of such models, to think a little more charitably about a *certain* kind of holistic hyper-connectivity. But Cramer quantum theory does not support mysticism, and Gribbin himself steals backward away from any such implication.

Davies, however, appears to be moving closer and closer to an explicitly black holist (although superbly informed) position. Here is a scientist in good standing whose books—when they are not called *The Forces of Nature* or *The Physics of Time Asymmetry*, by P. C. W. Davies—have eerily New Age titles: *The Cosmic Blueprint*, *The Edge of Infinity*, *God and the New Physics*, *The Mind of God*, and most recently *The Fifth Miracle*. These titles are not exactly misleading, since Davies has moved from atheism to a religious understanding of the world—though not the kind you'd be likely to hear at the local parish—via physics and now biology. Yet they are at least somewhat disingenuous, all the way to the bank. Dreaming up titles with this ambiguity makes a tasty parlour game: *Crystal Light Energy* (on lasers), *Maxwell's Demon!* (on thermodynamics), *The Shroud of Turing* (on artificial intelligence) ...

Consider these hedging remarks from *The Fifth Miracle: The Search for the Origin of Life* (1998), a title Davies presumably chose or at least authorised. 'I am not suggesting that the origin of life actually was a miracle'; 'Science rejects true miracles'; 'For many scientists, biological determinism is tantamount to a miracle in nature's clothing'.[28] But note the slipping and sliding. What Davies means by 'biological determinism' is not Creation Science, say, but traditional deity-free reductionism. It is hard to avoid the suspicion that this crypto-theological loading in his titles is what sucks in hundreds of thousands of readers who have never picked up a calculator for any task more gruelling than totting up the weekly bills.

RATIONALITY AND EMPIRICISM

Reductionism has been faulted for evaluating credibility solely in terms of rational and empirical criteria—logical reasoning, that is, and public, checkable evidence. One must not run the two together, of course. Any strict division between holist-introvert-intuitive and reductionist-extrovert-empiricist is vitiated exactly because each of us contains multitudes, and can shift modes depending on the demands of reality. It's like 'top down' versus 'bottom up'—both strategies are constantly in play, and often interact by way of Hofstadter's Strange Loops.

If it turned out (as some believe) that people empirically tended to be healthier and happier if they organised their lives around ludicrous cosmological tenets, perhaps you would have to grit your teeth and endorse the practical 'credibility' of these laughable or odious doctrines. If it turned out that you really could build a better engine or ecology by analysing them in terms that allowed internal contradictions—that is, by being 'non-rational', maintaining the truth of both A and not-A simultaneously—a preference for the empirical might oblige you to adopt those practices, despite the outrage to classical logic.

Actually, it is arguable that quantum theory, with its superposed and entangled states, *already* requires the abandonment of syllogistic canons of reason. And maybe pre-contact Aboriginal lifestyles exemplify the lifestyle benefits of governing your life according to the spirits. (But give me electric lights, CDs, high-speed dental drills and antibiotics any day.) Luckily, it seems likely that quantum complementarity is either restricted to its tiny realm, or will prove to be just a stopgap methodological fix until a still better physics is devised.

Besides, the social benefits of animism seem to me largely a by-product of the practical impotence of such doctrines. If your medicine is based on magic and your economy on foraging (however prodigious the memory and virtuoso skill required in its practice), your culture will simply fail to reach the population growth that, as its side effects, can produce pandemics and arms races and power hierarchies and exponential re-entrant knowledge growth. So among the benefits of believing incorrect things about the world is that you cannot build nuclear weapons but are restricted to such interventions as killing all the megafauna of an entire continent and thereby burning the original ecology to the ground over a few tens of thousands of years.[29]

Must materialists be reductionists? (Presumably not: Steven Rose, cited in this chapter's epigraph, is a dialectical materialist who repudiates reduction as a philosophical method, even if he adheres to it in the lab.) Well, this seems true at least: it is a materialist position—whatever that means in a quantal universe where matter is congealed force, or a relativistic one where matter is congealed geometry—to reject all hypotheses or methodologies that posit any top-down aspects of reality not derived by evolution from fundamental forces, particles or raw geometry. That still leaves lots of weird windows open.

It might be, for example, that human minds are, after all, somehow and in some respects quantally non-locally connected, as the parapsychology work I mentioned suggests. Or perhaps, to get really extreme, earlier intelligences arisen elsewhere in the galaxy have evolved by now to 'godlike' estate and secretly run the universe. Or maybe an end-of-eternity intelligence works backward through time to rearrange history as Sir Fred Hoyle claims. Or perhaps the phenomenal universe is actually and literally a cybersimulation in higher dimensional space, with us as subroutines ... If any of these were true, then a rigorous materialism would indeed require top-down explanations of many phenomena at our human realm (as the Gaia hypothesis might turn out to do, although James Lovelock's originating version is strictly materialist). But there's no strong evidence that these merry conceits actually have a real-world embodiment (except, maybe, those ambiguous data from the best parapsychology labs).

HOLISTIC VINDICTIVENESS

In practice, reductionist hypotheses would seem preferable wherever possible, just because that's the path science has followed to reach its goals in the past. Non-reductionist models (*élan vital*, for example, as an 'explanation' for life) have turned out to be notably misleading, based on bad analogies and wishful thinking. But materialisms, too, can be 'holistic' or non-reductionist, or claim to be; Marxism is only the most well known, although currently in eclipse. None of this has much bearing on issues of the mind, say, of the Freud-versus-Jung variety, except that as a materialist I become wary (or even furious) when I hear intelligent people airily invoking 'psychic energies' or 'ids' or 'Shadows' as if these metaphors possessed the same level of instrument-testable reality as electron flows and dopamine titres. Such figures of speech might have a metaphoric allure given our current comparative ignorance of the mind/brain-body, but they drag us into explanatory cul-de-sacs.

Watching *Passions of the Mind*, a series of television programs about the life and thought of Carl Jung, I was entertained in a horrified fashion by the way Jung's acolytes handled physicist Wolfgang Pauli's defection from the celebrated Jung-Pauli synchronicity hypothesis. (Black holists never tell us about that apostasy, oddly enough.) Pauli abandoned this great scientific truth, according to his former analyst, because he could not face the deep inner mysteries of the Shadow. These monsters therefore lay in horrid wait for him—and hence, she said with what seemed to me ill-disguised relish and self-satisfaction, he swiftly died of cancer. Admittedly, psychological stress can be shown (empirically and rationally) to obstruct immunological function via a recently discovered brain chemo-modulator pathway ... and so maybe stress hinders repair of early cancer cells. Nonetheless, the claim that Pauli got cancer and died *because he resisted Jungian analysis* is deeply offensive and self-serving. But quite typical of black holism's sometimes malicious or stupid magical thinking.

Well, perhaps that does not matter, if what counts is experiential richness of life. For millennia, wise people have obviously lived wonderfully satisfying lives without the slightest notion that the earth goes around the sun or that the blood circulates or that tiny invisible animals make you sick. But I cannot help thinking that to employ the best available physics and physiology has to be preferable to clinging to time-hallowed commonsense errors. It would be a terrible travesty of intellectual justice if our dawning era of complex, brain-stretching scientific solutions to the age-old problem of death became conscripted by black holists, warped into yet another comical New Age fantasy of effortless deliverance.

Gaia at Risk

Has science, as a common wistful tale suggests, driven us out of some golden Edwardian garden into a horrifying twenty-first century waste-land of feral technology, of which the most audacious emblem might prove to be the conquest of death? No. As ecological writer Michael Allaby has argued, it is our perception of the world—our core narra-tive—that changes. Currently it is 'profoundly pessimistic. We seem to find masochistic enjoyment in punishing ourselves.'[30] Nothing could be more masochistic than a supine embrace of mortality in an era when researchers have it on the run.

Allaby's collaborator James Lovelock is the inventor of the Gaia hypothesis that the entire world is, in some sense, a gigantic, self-regulating organism. Yet like Lovelock, Allaby is aghast at the mystical excrescences that have attached to this attractive scientific model. In particular, he reproves those who condemn 'reductive' science as wrong-headed or spiritually demeaning. Isn't the Gaia model of a living planet as 'holistic' as it gets? In a sense, but there's that paradox at the core of science. Computer modelling, for example, allows the study of very complex systems replete with feedback loops, including Gaia itself—'but it depends on an extreme reductionism,' Allaby notes. 'If the model is to reflect accurately what it represents, it must be constructed as an assembly of its parts, each of which must be known separately.'[31]

We long for a whole, unmediated world, but we must cut up our dinner in order to eat it.

six: cosmos – everything, & more

How to explain? How to describe? Even the omniscient viewpoint quails.
Vernor Vinge, *A Fire Upon the Deep*[1]

It was so old a ship—who knows, who knows?
—And yet so beautiful, I watched in vain
To see the mast burst open with a rose
And the whole deck put on its leaves again.
James Elroy Flecker, 'Old Ships'

Where *did* we come from? Where are we going?

To the cosmologist, the canonical answer to the first question is: from the Big Bang, some 15 thousand million years ago. Or maybe 20 thousand million. Or perhaps eight billion. That ballpark anyway. You can be pretty sure that 4004 BC is wrong.

Right now, however, there's still no agreed answer to the second question. Where is the cosmos headed? On the latest evidence, we inhabit an expanding universe that will stretch and thin without limit, cooling and attenuating until its dead stars and their decaying components pass into a heart-freezing eternity of darkness and silence.

Nobody loves this answer. We want a universe with closure. Better for the universe to perish in fire than in chill. Better it were returned to the kiln, in blazing crushed compression. Anything is preferable to uttermost void. We want a universe that might unfold again like a rose after the long night ...

So the specialists wriggle on their equations, seeking desperately for some trick that will make the vastness *thick enough*—perhaps with hot dark matter ('massive' but almost undetectably light and aloof nuclear particles called neutrinos, moving at close to the speed of light), or cold dark matter (dead suns, or brown lumps of gas that never quite ignited), or exotic particles with names like 'axion', or some mysterious novelty that will give the void enough ... *gravitas* ... to close like a fist at the end of time. In June 1998, it looked as if part of the puzzle was solved. Experiments by Japan's Super Kamiokande group, using a 50,000-ton tank of purified water ringed by 13,000 photomultiplier tubes, a kilometre below ground in the Kamioka Mining and Smelting Company Mozumi Mine, provided the first strong evidence that neutrinos, quite against standard opinion, do indeed possess mass—not much, hardly any at all, but perhaps enough to explain some of the mysteries of galaxy-scale gravitation.

THE HISS FROM THE BIRTH OF SPACE AND TIME

Two decades ago, the Nobel Prize for Physics went to Arno Penzias and Robert Wilson, two Bell Telephone electronics experts, for an accidental discovery that established the first answer (or so we currently believe): the origin of the universe. In what Paul Davies has called 'the greatest of all scientific discoveries', rather more than three decades back, when I was already a legal adult, these engineers detected a pervasive celestial radiation in the microwave range coming at us almost equally from all points of the heavens. This cosmic background radiation has a temperature, some 2.7 degrees above absolute zero. It is now understood as the relic of our explosively inflated beginnings, a trace from the ancient epoch when the ionised plasma veils of creation cleared a hundred thousand years after Moment Zero and the first stable atoms congealed into existence. With this critical evidence in hand, cosmologists swiftly pushed theories of the Big Bang back to the first one hundred thousandth of a billion billionth of a second, when all the mass and energy of the observable universe was compressed into the dimensions of a single atomic nucleus.

When I was a kid and no-one knew anything much, two of my most precious books told me about the origins of everything. (Three, if you count the Bible. At that stage 'Let there be light!' was almost as scientific as it got.) One of them had a wonderful, hubristic title, *The Creation of the Universe*, and turned out to be amazingly close to the truth—assuming we now know what that is. It was first published in 1952 by the unorthodox physicist and prankster George Gamow.[2] Its scientific and ideological rival was *Frontiers of Astronomy*, published three years later by the equally eccentric astronomer Fred Hoyle.[3]

Hoyle twisted my youthful mind with a silly, wonderful question: *Why is it dark at night, dark between the stars?* After all, if the universe extended without limit—and there was no reason to think it didn't— there should always be further bunches of stars (no matter how faint) filling in the stipple we actually see, until the whole sky blazed like the sun. Hoyle's answer drew upon astronomer Edwin Hubble's majestic insight of the 1920s: *The sky is dark at night because the Universe expands.* (Yes, the same Hubble whom the Space Telescope memorialises.) The galaxies are fleeing from each other as spacetime stretches endlessly, opening voids of blackness that will never be filled with light unless the cosmos closes once again.

But if that were so, as for a whole human lifetime most cosmologists have now accepted, maybe it was expanding from a single moment of origin. Its likely fate appeared to be sealed, if the Big Bang proved to be a one-way process, a flight to forever. Then the scattered parts of the universe would never return, summoned by gravity, into a final state of recapitulated infinite compression. No, the last stars would flicker out in a hundred billion years' time. Many would be swallowed into black holes. For a billion billion (10^{18}) years, space would expand in lifeless blackness. Finally, 'after no less than ten thousand billion billion billion billion billion billion billion billion years' in the frozen wastes (according to a calculation by Davies) the decaying black holes would begin radiating like ... fireflies. Feebly, that is, in the utter darkness. After a further million billion billion billion years, 'as their temperatures climb more rapidly, they will disappear in a bright flash of radiation' which will ebb endlessly in the ultimate void.

It was an extraordinary story to piece together from a handful of equations and a handful of meter readings, especially its origins in a bursting flare from nothingness. Might not an infinite, eternal universe be preferable, a cosmos according to Hoyle? Ignition, outburst, a first moment of space and time from the nothing, the nowhere, the no-when: that was Gamow's claim, and he was humble enough in the 1950s to admit that it was 'probably too early to say which of the two points of view will ultimately prove to be correct'. He did mention, though, that such a cosmic explosion would leave a trace: the freezy embers of a universal fireball, some 50 degrees, he expected, above absolute zero. Not right, but close enough.[4]

Neither of these wonderful, child-bending books used the term 'Big Bang', because Hoyle hadn't yet coined it as a sarcastic put-down of Gamow's theory of the 'big squeeze'. Hoyle's own notion was the 'steady state', in which matter and energy expanded endlessly into infinity, continuously replenished by an unknown force-field. In less than a decade, though, he was proved wrong, when that remnant background radiation was detected. It was even cooler than Gamow had predicted, by more than 40 degrees (he'd made a mistake in his add-ups).

Science is the perfect consumer item, endlessly shifting last year's stock into the junkyard. Harvard physicist David Layzer has proposed a different story again, a highly random *cold* origin, distinctly unfashionable but intriguing.[5] And the maverick commentator Eric Lerner boldly declares that 'the Big Bang never happened', pinning his own model of a universe infinite in time and space on the magnetic plasma theories of Swedish physicist Hannes Alfvén.[6]

Still, following scientists on their great collective enterprise, their intellectual journey from the Big Bang to the nature of brain and consciousness, proves that to travel hopefully is better (and more stimulating) than to arrive. At least until immortal humans finally, truly arrive at an unimpeachable conclusion.

STRETCH MARKS ON THE SKY'S BELLY

While the standard story of creation has a highly ordered fireball Big Bang, inflating instantly to a smooth expanding cosmos, it can't easily explain why galaxies clump into vast walls hugging empty cosmic bubbles, a bizarre feature only fairly recently discovered by astronomers. Ultimately, order emerges in hierarchies of self-organisation, attributable in the final instance to the very expansion of spacetime itself. In recent decades, it's true, the Big Bang theory has been improved by the theory of inflation (who says scientists are out of touch with economic reality?), which welded cosmology with quantum theory—the absurdly large and the ridiculously small. But problems remained. Above all, the layered structure of the visible universe went unexplained. An inflationary Big Bang, bursting instantly from zero to literally cosmic proportions, should have left spacetime so smooth and featureless that no galaxies, no stars, no people could evolve. Unless ... tiny quantum imperfections at that initial fraction of a second were also inflated, creating vast 'wrinkles' or ripples in spacetime, the irritants upon which matter might precipitate the pearls we call stars.

George Smoot, a researcher at the Lawrence Berkeley Laboratory in California, together with literally hundreds of colleagues, worked for years to map the faint tracery of those ripples in the sky, and in 1992 announced their discovery, to tremendous if fleeting press enthusiasm. In effect, their arcane efforts had confirmed Gamow's guess, and in turn seem to have been confirmed by even more recent work in space and from the ground. So we now knew with some confidence how the universe was formed (although Hoyle was *still* not convinced, holding out for a blend of general steadiness and local Bangs). Smoot thought his computer-enhanced pictures were like looking at the Face of God.[7] To me, it's more like the beaming grin of George Gamow. Suddenly, unlike the dreamy cosmologies concocted beside the embers through a hundred thousand years of human wondering, we possessed a creation story, a cosmos, that could stand up in court. We knew how the universe was built! We had seen the traces of its wrinkled newborn torso, wrenched in its birth from that first and greatest of cosmic singularities.

A Theory of Everything, Finally

And we had some good hopes of knowing why the universe took the shape it did. Paul Davies had said as much a decade earlier:

> [W]e can now glimpse what a complete theory of all existence is like. We can see how such a theory is possible. We can at last comprehend a universe free of all supernatural input, a universe that is completely the product of natural laws accessible to science, yet which possesses a unity and harmony that manifests insistently a strong sense of purpose ... [T]he unification of physics has leapt forward, and the outline of a complete theory of nature can at last be dimly perceived ... [F]or the first time in the history of science we can form a conception of what a complete scientific theory of the world will look like.[8]

By the close of the millennium, that hope was edging closer, perhaps on schedule. Madhusree Mukerjee, writing in *Scientific American* in 1996, told a similar story to Davies' (even if the details, depending on a second generation of arcane string theories, differed somewhat): 'A sense of barely suppressed excitement fills the air. The Theory of Everything, or TOE, the theorists believe, is hovering right around the corner.'

A *complete* scientific theory of the world, hovering right around the corner! Hubris enough to curdle the sweet milk of Creation! It was worth heeding the more measured remarks of British astronomer John D. Barrow, offered in 1998 in his ominously titled *Impossibility*: 'Ask the elementary particle physicists what the world is like and they may well tell you it is very simple—if only you look at it in the 'right' way. Everything is governed by a small number of fundamental particles. But ask the same question of biologists or condensed-state physicists and they will tell you that the world is very complicated, asymmetrical, and haphazard.'[9]

At a deep level, it's true, there is no inconsistency in these two kinds of comment. It *does* now look as if the universe is built of only a few kinds of basic bits and pieces, or fields, or geometries. When those bits are multiplied into the unspeakable trillions of trillions of trillions, and they are let loose to play with each other during billions of years, we end up with a universe that is very ornate and thorny indeed. Even if mathematical physicists do attain a theory of such elegance and parsimony that it literally accounts for all the elementary components and forces of the cosmos, a theory so powerful and simple that it authentically deserves the name Theory of Everything (Basic Model), it is important to grasp, as Barrow adds, that this verges on a misuse of ordinary language.

> To the outsider 'everything' means what it says—everything, with nothing left out! But this is not what physicists mean ... It is not an oracle which will print out an explanation of everything we see in the Universe, together with a list of all the other things we could see if we looked in the right place.[10]

This sounds properly modest. But watch carefully. Using nothing more than the mathematics of general relativity and current elementary particle physics, the canonical Big Bang is about to be performed. Before your very eyes, I will conjure the universe.

Out of the Nowhere

Hat is empty. Hands are empty. Here comes the rabbit! Applause, please. But wait! There was no hat. There were no hands. The rabbit popped out, believe it or not, from the purest vacuum. But hold on: even the vacuum wasn't empty. No, the vacuum seethed and roared with quantum 'virtual particles', fleeting electrons and positrons, quarks and anti-quarks, a huge family of fundamental particles, bounding from the Void like rainbow-sprayed dolphins, crashing back into annihilative mutual embrace before you could get a fix on them. O Heisenberg!

So seeing the rabbit hurtle from the vacuum is not so impressive after all. Go back a step. Watch, watch. Here comes the vacuum itself. But it's not the one we asked for! It's a false vacuum, choked with impalpable energy, teetering on a metastable lip of nothingness. It trembles, it slips, it falls: false vacuum slams catastrophically to the bottom of the graph, spilling its prodigious force into the true vacuum it creates and inflates, blowing out like a balloon hitched to the hot end of a Jumbo's motor, a whole skyful of balloons going from nothing to something in one metaphysical jump to freedom, doubling and redoubling in fractions of infinitesimal fractions of the First Second of Creation, 50 or 70 orders of magnitude of endlessly stretching space in an interval so small our minds cannot even begin to comprehend it without equations to hold it at a decent distance.

So here's the rabbit after all, here comes spacetime out of the genuine nowhere, the nothing-at-all which preceded the Planck Instant, that very first 10^{-43} second of now-ticking time, when all there is was jammed into a region less than 10^{-50} of a centimetre across. At that earliest time the four known forces of physics (gravity; the colour force that holds quarks together at the heart of nuclei; the 'weak' force that bursts nuclei open; the electromagnetic force that builds atoms into molecules) were, all four, submerged into one sublime unitary bond. Then the colossal expansion of the new universe sucked away the heat of creation (much as vented steam, expanding, cools into dew) and cracked today's four forces apart, one after another, in a domino tumble of 'broken symmetries'.

In such a cosmos, finally understandable to the human mind (maybe, with luck), what is the status of that mind, that cosmos? Steven Weinberg, the 1979 Nobel Physics laureate whose insights produced much of our current grasp of the moment of genesis, is notoriously bleak: 'The more the universe seems comprehensible, the more it also seems pointless.'[11] Gaze toward the shutdown of the universe, the final Whimper, and certainly it looks as if the Copernican dethronement of human values is now complete.

THE END

For, yes, the same mathematics and astronomical data that yield a glimpse into the origin of things show us also the end. A variety of ends rather, as noted earlier, for it is still not clear yet if the universe is due to falter and collapse back into the chaotic inferno of a Big Crunch, or expand without limit into frigid vacancy (the current best bet, based on Hubble evidence). It may even be that the very vacuum in which our universe is embedded is not stable—not yet at the lowest state it could conceivably adopt. If so, it might decay further, and a bubble of true vacuum could even now be roaring toward us at the speed of light gobbling up all the matter and energy in its path. Fans of instant apocalypse will embrace this notion with delight. Luckily, there's no evidence in its favour.

Indeed, the latest suggestion is that cosmology might stand at the edge of a total revolution, driven by new information flooding in from advanced instruments. While the vacuum energy of the cosmos might not be unstable in the way just mentioned, its value certainly might be different from any currently accepted. For a start, the fact that new data seems to show that the universe is actually *accelerating* its expansion is not easily reconciled with the standard models. Peter Coles, writing in *Nature* in June 1998, remarked: 'For many of us, that is the most exciting possibility of all, as we would have to move to stranger theories, perhaps not even based on General Relativity.' And in October 1998, calculations by Charley Lineweaver were reported in *Astrophysical Journal Letters*, based on detailed observations of the cosmic microwave background 'wrinkles' in spacetime, together with data from constant luminosity supernovae of Type 1a, galaxy cluster mass-to-light ratios, and double-lobed radio sources. These findings, woven together, indicated that perhaps 25 per cent of the cosmos is comprised of dark matter, while a whopping 70 per cent of all energy is locked up in the vacuum itself (more strictly, the value of the cosmological constant is 0.7), providing the unstoppable acceleration.[12]

All this, needless to say, will be of considerable interest to any intelligences still thinking about such matters after immortality is secured, because it is possible that they themselves will *still be around* (or their multiply-modified, expanded, uploaded or cloned 'selves') when time at last runs out, runs away, or ebbs to a standstill.

Certainly our Sun will be gone in five billion years—a length of time equal, curiously enough, to the span since its birth. In its death pangs, expanding to a red giant that nearly swallows our planet, it will boil off the atmosphere and oceans and sear the surface, melting everything natural and constructed. In a trillion years, *all* the stars will be dead of natural causes. It is a prediction of versions of grand unification that in 10^{37} years from now all the nuclei in all the atoms in the cosmos will have undergone spontaneous disintegration. By 10^{44} years, aloof hypothetical axion particles would have decayed into photons. Even matter trapped inside fabulous black holes that have swallowed entire clusters and superclusters of galaxies will evaporate, given enough time, into almost infinitely dispersed radiation remnants: 10^{117} years (to put a neat figure on it: hubris!). The quasar-capped galaxies, the gorgeous palaces of time, the solemn temples, the great Cosmos itself, yea, all which it inherit, shall dissolve. Leave not a rack behind.

THE SECRET CODE OF THE COSMOS

If this scientific perspective remains bitterly strange to the liberal arts mind (despite its own bouts of angst-ridden existentialism and deconstruction), alien to the instinct of the historian or the literary theorist, that's because it can only be rendered truly in the tints and tones of logic and mathematics.

For those of us delighted by ideas, even topics that frighten many non-scientists can become entertainments. Mathematics is the most notorious. Clearly nature has numerical underpinnings. Proofs, as mathematician Ian Stewart has pointed out, are usually presented as dreary slogs of logic, unconnected to anything in art or our ordinary lives. But 'both a proof and a novel must tell an interesting story,' Stewart observes; 'A good story line is the most important feature of all'[13]— which is one reason why mathematics can be a passion as well as a tool for building enormous bridges and telescopes in space. Ultimately 'we must develop a new kind of mathematics,' as Stewart says, 'one that deals with patterns as patterns and not just as accidental consequences of fine-scale interactions', where physics—even the physics of creation—tends to remain.[14]

Mathematics has always been a cornerstone of classical education. While there are strong pragmatic reasons for retaining mathematics as a crucial aspect of everyone's learning, the sad reality is that most of us are gobsmacked by anything harder than simple add-ups (and, as we've seen, even bright guys like Gamow can get calculations wrong). Business spreadsheets zapping away on the office computer burst some of the computational barriers, and pocket calculators spare us the rote task of learning our multiplication tables, but mathematics itself is as fraught a ever. Still, the central importance of mathematics in the contemporary demarcation of science from other forms of discourse and practice is emphatic. If a body of observations cannot be ordered by a mathematical model, it is highly unlikely to be accepted in the workshop of science. Perhaps 'workshop' is the wrong word. 'Fortress' catches it better, for the boundaries of science are patrolled by a self-ordaining interpretative community of practitioners (and a good thing too, by and large).

Those 'soft' disciplines which lack such foundational formalisms as time-evolving equations like the Hamiltonian (a basic mathematical gadget connecting energy and rates of change, in the calculus) tend to be permitted entry only to the lobby of the workshop, as defined by its high-prestige members, the mathematicians, physicists and maybe chemists. 'To the outsider,' Paul Davies has remarked, 'mathematics is a strange, abstract world of horrendous technicality, full of weird symbols and complicated procedures, an impenetrable language and a black art.' But he adds tellingly: 'No one who is closed off from mathematics can ever grasp the full significance of the natural order that is woven so deeply into the fabric of physical reality.'[15]

The root of the problem is the origins of mathematics in simple counting. Even people who find maths impossibly alien assume that counting is an innate human skill. Yet many cultures know only one, two and 'many'. The power to enumerate quantities larger than that, using chunks of fives and tens that we can map onto our fingers, is a major breakthrough. So if simple counting is itself a peculiar skill, what are the sources of mathematical truths? Are they dug out of the world, like other practical discoveries? Are they invented, never having existed until they are thought by a human mind? Or do they reflect an unchanging Platonic realm transcending space and time? Oddly enough, there has been a resurgence of popularity for this latter opinion. Do the laws of physics ordain what arithmetic and geometry we can do, or as John Barrow has asked, are they 'themselves consequences of some deeper, simpler rules of step by step computation'?[16] This curious possibility encourages a new kind of experimental mathematics, endlessly open, yet strangely ugly to those traditionalists for whom both mathematics and physics are a kind of sublime poetry.

Like the heart of a poem, the meaning of a scientific theory eludes rude paraphrase or translation. So we are left angrily chafing at the absurdity, all these decades after the luminiferous aether skulked off without a pension, of the cosmos inflating itself out of a ... well, it's not my fault ... a *false* vacuum.

Wild hubris? Perhaps the second-hand pride one feels at these prodigious contortions, as pure thought wrestles brute reality to the mat, needs to be chastened by a deeper realisation that the underpinnings of the human mind are, after all, matched in advance by evolution to the shape of the universe. Surely it need not have been that way.

WHAT SEX IS SCIENCE?

Then again, perhaps it has not been. Perhaps our belief that it is merely reflects local prejudice—culturally bound, racist, sexist.

Consider Pythagoras as an icon of this contest. A mystic who flourished 2500 years ago, he founded an order of devoted mathematicians who disdained the Greek fashion of frocks for manly men and took to wearing Persian trousers. Men have been wearing the pants in physics ever since. Girls don't do science, do they? And surely they hate mathematics. 'Being a "brain" is all too often seen by teenagers of both sexes as unfeminine,' according to Margaret Wertheim, a science writer with degrees in physics and mathematics. Hence, 'many bright girls take the more socially acceptable path of pretending to be less intelligent than they are.'[17] (Recently, of course, schoolgirls have out-performed boys in these very subjects.)

As a result, according to Wertheim, physics has become the domain of 'Mathematical Man', with his 'almost antihuman focus, deeply alienating'. Luckily, 'the problem is not *that* physicists use mathematics to describe the world, but rather *how* they have used it, and to what *ends*.'[18] Ends such as, precisely, the search for the Theory of Everything, a 'socially irresponsible' quest that drives this new priesthood to squander billions in raising 'ever more elaborate cathedrals' with results that are 'not only utterly irrelevant to daily human life and concerns, but also incomprehensible' to those who have taken the socially acceptable path of feigned stupidity.

The 'cathedrals' that appal Wertheim are not the 50,000 nuclear weapons that still exist in an unstable world, nor the reckless industrial programs threatening the world's wellbeing. They are the huge particle accelerators that slam protons and other subatomic components together at nearly the speed of light, sucking new kinds of matter out of prodigious raw released energy, testing the models of order and chaos that physics offers. The impulse behind these Promethean efforts has been explained by Stephen Hawking and others, in a catchpenny stroke of PR rhetoric, as seeking to know the Mind of God. Hawking, in reality, is an atheist. He means nothing more (but certainly nothing less) than this: seeking the deepest rules of space and time, geometry and energy, the most succinct compression into neat equations of all the forces and fields and particles making up the cosmos.

Is this publicity gambit—which persuaded millions to buy Hawking's difficult book—a valuable clue to secret religious motives behind today's physics? Feminist critics of science tend to regard it as a male priesthood that, like other monasticisms, systematically excludes women, or makes life very hard for them if they do manage to get in the door. And you know what men are like. To fix things up, Wertheim suggests, we need to 'ground physics in an ethical and socially responsible framework'.[19] Mathematical Man has to 'embrace the partnership of Mathematical Woman'. Maybe so. The horror stories of injustice to nascent women physicists are harrowing and abundant. Lise Meitner, who worked out the theory of nuclear fission with her nephew Otto Frisch, was ignored when the relevant Nobel Prize went to another Meitner collaborator, Otto Hahn, in 1944. But then, so too was Frisch. And it's true that Meitner's first choice as a research supervisor had turned her down—yet this was not a bigoted man, but the celebrated Marie Curie. Meitner's boss Emil Fischer would not allow women into his chemical institute, so Lise had to work in the basement and hide under the seats in the lecture theatre to audit courses. Fischer, though, came good when Prussian policy changed, 'even installed a separate toilet for her, and eventually became one of her greatest supporters'.[20]

So male enmity for would-be women physicists is not universal, nor has it worked as effectively as the Catholic Church—Wertheim's analogy—in excluding women from its priesthood. Indeed, as she shows, women earn 47 per cent of US degrees in mathematics and statistics, if only 15 per cent in physics. Women earn a fifth of the math/stats PhDs, but just 11 per cent in physics. Still, from a standing start at the beginning of the century, these are really quite encouraging figures.

Would physics be altered drastically—made more touchy-feely, perhaps, and socially redemptive—by an infusion of clever women? With nervous hindsight, we might doubt it. The distinctive female genius of Meitner and Curie, after all, helped lay the foundations for Hiroshima and Chernobyl. This sinister link need not be taken too seriously, of course. Nobody knew the convulsive possibilities lurking inside the radioactive atom, so we can hardly blame the women researchers. By the same token, no-one yet knows what benefits and disasters will unfold from a truly successful Theory of Everything (we might trigger a vacuum catastrophe, for instance).

But in any case, why *should* we interest ourselves, and at public expense, in the life and death of distant stars, or of the universe entire? Why, indeed, should we care about science at all?

SCIENCE IS HUMAN

Wertheim's version of male Science makes it as remote and cold as the stars themselves, turning inexorably over our heads, the uncaring machineries of night. Wrong, and not just because stars are hot rather than cold. For over 300 years, science has remade our world by treating it, including the people who benefit from this mixed blessing, like a machine. According to many modish science writers, as we saw in the last chapter, the scientific revolutions of this century (relativity, quantum theory, chaos, complexity) superannuate such bleak reduction. For twenty-first-century science, Paul Davies and John Gribbin have assured us, it will be correct 'to dismiss the notion of the ghost in the machine— not because there is no ghost, but because there is no machine'. I find this claim difficult to take seriously. To repeat: reductionism's excessive claims have been amply demythologised in the last 30 years—Arthur Koestler, now scorned or forgotten, started the ball rolling[21]—but it's still the only effective scientific game in town. I believe you'll find that this book you are reading, incidentally, was printed on a machine, not a ghost.

As we saw earlier, quantum physics, with its indeterminacies and leaps, does seem utterly disjunct from the commonsense world we construct at our household level of Newtonian mechanics. So does the unutterably vast cosmological landscape—a universe eight or maybe 15 or 18 thousand million years old, spread out as many light-years on either hand. Yet the human mind can generate equations which fit that cosmos with the same dexterity with which it generates poetic and practical writing. The world, then, is presumably—at some profound, non-obvious level of abstraction—all of a piece. What a stroke of luck! The rules genetically selected into our nervous systems by the requirements of language use, ecological management (formerly Hunting and Gathering), not to mention swinging from branch to branch without smashing our skulls, derive from the same underlying consistencies which shape the isotopic spin of atoms, the turbulence of rivers, the glacial rotation of galaxies, the billion-light-year filamentary streamers of galactic super-clusters.

Is Life Inevitable or Nearly Impossible?

Life, we once supposed, coalesced in a warm pond or thin organic soupy ocean on the early earth, perhaps 3.8 billion years ago. Now that seems unlikely, and the new prime habitats are infernal or celestial. Life might have started deep beneath a crust relentlessly bombarded by ruinous debris from space—or fallen from the heavens, in comet-borne spores or meteorites hurled up from the surface of Mars, then more hospitable to life than our own poor smashed planet.

If life formed by sheer accident, Paul Davies has argued recently, that in itself is a statistical miracle of biblical proportions. And if it happened because the universe is congenial to life, then something wonderful awaits us in the deepest equations: 'a self-organizing and self-complexifying universe, governed by ingenious laws that encourage matter to evolve towards life and consciousness. A universe in which the emergence of thinking beings is a fundamental and integral part of the overall scheme of things. A universe in which we are not alone.'[22]

Not alone? Davies means alien intelligence is likely elsewhere in the cosmos, emerged without miracles and developed into its conscious estate by Darwinian selection. But one would be excused at seeing our reprieve from loneliness (reading between the lines only too easily) as the Hand if not the Mind of God, reaching down to cup us in our apparent contingency, our cosmic lack of meaning. Most non-scientific readers, I suspect, will take spiritual succour from such words.

Davies' own solution emerges at the very boundaries of current theory. As well as spacetime and mass-energy, he argues, information itself must be fundamental, creating altogether distinctive kinds of causation until now overlooked by reductive science. This means 'accepting that information is a genuine physical quantity that can be traded by "informational forces" in the same way that matter can be moved around by physical forces'.[23] It is a bold proposal, echoed in papers only now emerging from quantum theorists and experimenters. If it is correct, life could be everywhere, awaiting the deathless humans of coming centuries as we move outward into the galaxy.

Are We Alone?

It might not be that way. Frank Tipler has tried to establish the rather old-fashioned notion that humans are alone in the universe. Destiny beckons not to *us* but to our silicon descendants, the independently intelligent automata we might choose to launch into the heavens as our ageless deputies.[24]

In brutal precis, here's the argument: any culture with an ounce of grit and curiosity will build at least one of these artificially intelligent, self-replicating star-probes. After a few million years, by natural increase, the galaxy will be rife with its offspring. Since a good proportion of the many fecund worlds predicted by optimists like the late Carl Sagan must be a few million years ahead of Earth, and since it takes only one culture with gumption in a universe of wimps to get the ball rolling, we should see such a probe in the sky. We don't; ergo we're alone in the galaxy (or at the very least ahead of the pack).

So searching for extraterrestrial intelligence may be a waste of effort and resources. If there is smart life elsewhere in the galaxy, you might doubt that it would be able to hide its presence. Beyond the twenty-first century, even on earth, the picture gets hazy, as science starts to look like magic. Ultimately, spacetime remodelling might involve what have been categorised as Type III or IV galactic civilisations, draining stars for energy and looping time into pretzels. (Here and now, we are still stuck at the border of Types 0 and I.) One plausible extrapolation is toward truly gargantuan works of engineering: stars mined for anti-matter, changed in their courses, ignited to supernovae and snuffed out so that their neutronium cores might be used for massive data storage. So unless we are the first kids on the technology block (an absurdly unlikely chance if life is widespread), it is a good bet that we are alone— at least in this galaxy of 200 billion stars. Which implies an awesome responsibility. Perhaps we are the fluttering candle of all future life in the universe. The sooner we share the spark the better.

Turtles All The Way Up

Physicist Lee Smolin has suggested that black holes extrude out of our universe and expand as Bangs into new universes, remixing the laws of nature a tad each time, a line of thought similar to Hawking's recent thinking about baby universes linked to cosmic thinking.[25] Smolin's story involves infinities of time, cosmos following cosmos, universes giving birth to baby universes that Bang and expand and fade into attenuated living death or perhaps are crushed into fiery nothing—but not before giving birth to billions of new baby spacetimes of their own.

The core idea stems from the mathematical discovery, crucial to quantum theory, that the vacuum—emptiness—is unstable and liable to emit energetic particles. What's more, the briefer this catastrophe, the greater the available energy. At high enough energies, this flaw in emptiness can buckle inward to form a black hole, which may then inflate wildly to form an entire segregated cosmos. Depending on the interaction laws coded into that cosmos, it might recollapse in a tiny fraction of a second, bubble with micro black holes that collapse in turn, or even blow out into a fresh spacetime universe akin to our own.

Imagine, then, a fabulous series of popping bubbles in the void. Most vanish with no progeny. Some persist long enough for their colossal outpouring of energy to form elemental vibrating strings and membranes, and for those to coalesce into quarks and electrons, and then into hydrogen and helium atoms. If the values describing the new universe are suitable, its contents may form stars that cook new, heavier elements in their compressed gasses. Some of these stars, rich in carbon and other key elements, will explode in thermonuclear glory, their cores crushed down to black holes, their hot debris seeding the skies with the raw materials of life and new stars.

In a few universes, Smolin speculates, an extraordinary set of values will coincide, yielding two interesting features: a maximum number of stars suitable for making black holes (and hence new universes with slightly variant parameters) and many other stars and their planets suitable for evolving life, and even intelligence. The key step in this somewhat Darwinian argument is that the parameters determining the shape of the new cosmos must resemble those of its parent, rather than being picked at random. Granted this, after a while most universes in the megacosmos will cluster around certain apparently arbitrary values—the kind that produce, by sheer coincidence, carbon-based observers like us.

By remixing the fundamental parameters a little each time, eventually (by a process more akin to Gerald Edelman's 'neural Darwinism' than to genuine Darwinism, in which phenotypes compete in the same environment for air time) certain kinds of fecund universes will dominate. As a side-consequence, it might be that these kinds are hospitable to life. Suppose a set of laws congenial to the manufacture of abundant black holes will birth more universes just like mother. Barren ensembles of laws will die out. After a while, most of the universes in superspace will be fertile. Smolin believes ours is just that kind. Intriguingly, most black holes are made when very large stars explode as supernovae, which also happen to build the elements of life out of the simpler elements originally forming those stars. So it's a side effect of universes that make oodles of black holes that they need to hang about long enough for such stars to form, to explode, to seed the cosmic dust with the making of life, which will then evolve and start wondering about the meaning of life. And the beauty of Smolin's suggestion is that it's testable. It moves out of the domain of faith and into that of science.

A more extreme possibility is that once life arises by chance and thrives, it will thereafter manipulate the parameters in one or more of the baby universes such that life-supportive universes (perhaps available for colonisation via wormholes) become prevalent. This wild notion has been explored by cosmologist Edward Hamilton in the sombre pages of the *Quarterly Journal* of the Royal Astronomical Society.

I am attracted to Smolin's view of a Darwinian meta-universe replete with 'gosh numbers' (as Frederik Pohl calls the relevant 'dimensionless' values which describe reality's architecture) selected by the fecundity of their local bubbles. However, he not only gives no mechanism for mutating the fundamental values, but provides no hint of why they should be constrained. Neither of the Darwinian prerequisites for natural selection are present. There's no reason to suppose that systematic and heritable variations would occur between bangs (rather than none at all or, more plausibly, wildly divergent ones), and there's no competition for resources. You can start a baby universe just by goosing an infinitesimal wormhole. Cosmic inflation does the rest. Pure science! Pure magic!

Recall the logic of organic evolution. Suppose that the earth had been an infinite flat plain, or plane, in which various regions were connected by hyperspatial wormholes. Then the original pre-DNA self-bootstrapping critter might have spread through the ooze forever, chomping and mutating and dividing as it went. But there'd be no concerted contest to drive evolution, so everything (it seems to me) might stay pretty much at slime level, unless complexity theory and such controversial factors kick in and smarten things up eventually. Smolin's story gives no mechanism for mutating the fundamental values, nor does it provide a hint of why they should be constrained. Stars allegedly evolve 'to make best use,' as science writer John Gribbin summarises the theory, 'of the available materials'.[26] But there is no coding transmitted between stellar generations and no memory substrate to mutate—just a shift in the population of raw elements from previous supernovae generations.

Watching galactic spiral arms, Gribbin has suggested, it is obvious you're seeing a living system. By this reasoning, the water vortex I saw tonight in my bath was also alive, poor little thing. 'Many universes,' he claims, 'are competing with one another for the right to exist'. But the mathematics of this story tells us the universes are *orthogonal*, at right angles to each other, so to speak. So it is difficult to see how they can 'compete'. And for what resource?[27]

BEYOND MIRACLES

Perhaps the most striking feature of the crypto-religious dispute, though, is the way science and theology continue to skid past each other. It is 60 years since Karl Barth taught Christians the 'wholly otherness of God'. Just as scientists such as Davies arrive to marshal arguments for a Necessary Being who can bind the sweet influences of the Pleiades, or loose the bands of Orion, deconstructive theorists like the brilliant poet and academic Kevin Hart are blending Heidegger, Derrida and the pseudo-Dionysus in dazzling feats of negative theology. The Way that can be named is not the Way, no way. Religious faith by definition evades rational grasp. Davies, a deist rather than a Christian theist, nonetheless has his most ardent admirers among those who hold with the entirely supra-rational first article of Barth's 1934 Barmen Declaration: 'Jesus Christ, as He is attested for us in the Holy Scripture, is the one Word of God which we have to hear and which we have to trust and obey in life and in death.' You might suppose that such a naturalistic quest risks ending in the bathos which closed John Updike's novel *Roger's Version*, a mathematical computer search for the face of Yahweh. Updike saw this marvel on his own word processor: 'A mournful countenance gazing out through the scrambled numbers, a squared-off, green-on-green Veronica.' It is time for us to move on from this divine dinosaur, this god out of the mathematical lexicon, this Roger's Theosaurus.

UNWEAVING THE FABRIC OF REALITY

Yet another alternative view has been provided recently by David Deutsch, quantum theorist at the Clarendon Laboratory, Oxford University.[28] His shocking idea is even more resistant to the mind's clutch. We have glanced at more primitive versions of this idea already, usually known as the Many Worlds model. Deutsch sees every event literally doubled and reduplicated, with small crucial variations, in trillions upon trillions of diverging parallel realities, spread through infinite lateral time. If you're having a bad hair day, there is another version of you, in a universe at right angles to this one, who's doing just fine. And a zillion more, taking up every conceivable alternative position in between. Some of them are dead. Some are on Mars. A few are sharing Graceland with Elvis, who is married to Princess Di or perhaps John Lennon.

Neither of these remarkable postulates is freshly minted by Deutsch or Smolin. Baby universes go back more than a century to Charles Peirce, semiologist and pragmatist, while the Many Worlds hypothesis was proposed in 1957 by Hugh Everett, III, who in this universe died decades ago. Until recently, though, nobody had pursued these striking notions with such ruthless intellectual attack, following them all the way down.

Deutsch, curiously, insists that his outrageous version of reality ought not surprise us, or at least physicists, since it is simply the best theory available to science (quantum mechanics) taken perfectly literally without metaphysical evasions. He argues brilliantly that something as ordinary as the splash of illumination shone by an electric torch on a wall cannot be explained without acknowledging that a sort of 'shadow light' from many other adjacent universes is leaking into our own, interfering with the beam we see and blurring its bright sharp disk into concentric rings.

In those universes, naturally, a few photons from our local universe also leak across, fritzing *their* beams of light. 'It follows,' Deutsch notes, 'that reality is a much bigger thing than it seems, and most of it is invisible.'[29] Even so, the multitude of overlapping histories are needed to explain some of the most basic features of our ordinary world. Deutsch argues ingeniously that gene sequences performing the same function in different adjacent worlds must be closely similar, while so-called 'junk DNA' littering genomes will vary at random. It is precisely this consistency in true informational structures that marks them off from noisy rubbish, however complex and elaborate the noise might seem.

For Deutsch, this truly weird interpretation of basic science lies at the heart of an emerging fourfold Theory of Everything Knowable (which is only a small part of Everything). It will compress our understanding of evolution, theories of knowledge and quantum, and computation. These strands seem very disparate, but he argues powerfully that where they combine we find an account of reality's fabric that escapes the mechanistic or Newtonian character of standard reductionist science. Deutsch, one of the founders of quantum computing, is no New Ager. His extravagant ideas are worth taking very seriously indeed.

ALL THE INFINITIES

Can the two new infinities—endless levels of cosmic histories, incalculable overlaps of variant worlds in the 'multiverse'—meet at a single point? Deutsch told me his own model 'is completely consistent' with Smolin's cosmological theory. 'These two types of multiplicity are completely independent (at least, according to our best existing theories).' Doesn't his claim that we each live every possible history obliterate morality, as I argued earlier? If my 'free' choice of a good deed requires that at least one version of me has taken the 'evil' fork—not just that this was in some abstract sense 'possible', but has actually done the deed—then Mother Teresa was, somewhere in the multiverse, a villain. This is not a physical objection to his account, but it is a distinctly distressing one. Deutsch disagreed: 'We have to re-learn how to react emotionally to what happens in other universes.' He added: 'This doesn't mean that events in other universes have no moral, or emotional value in ours. The weight which we should attach to them is given by the theory of probability, which also only makes sense in a multiverse context, but which,' he wrote, no doubt with a grin, recalling the mathematician Fermat's remark that has puzzled others for centuries, 'this margin is too small to contain.' Presumably he has already solved the matter of multidimensional ethics ... in another universe.

Yet another (apparently) drastically spendthrift theory is the mathematical model devised by Max Tegmark, a fellow of the prestigious Institute for Advanced Study at Princeton. His candidate for 'ultimate ensemble' Theory of Everything has a distinctly Platonic cast, for he argues that every conceivable mix of dimensions, constants and so forth comprises the multiverse. He even provides a chart of these mad universes on his Internet site, illustrating quite compellingly his thesis that only a tiny intersection of all the basic parameters can sustain life—'observers'.[30] We fit snugly in that box.

How can we know this, when the entire argument is about realms of being that in principle we are excluded from contacting? As usual, the reasoning is abstract, but powerful because it derives from the deepest theories of science. It boils down to this: a cosmos with no space or time is presumably empty. And however many time dimensions you allow a universe, if it has no spatial extension 'physics has no predictive power for an observer'. The same holds for the obverse case with one or more spatial dimensions but no time. (Technically, these are built on elliptical partial differential equations.) With one dimension of space, and more than four of time, the worlds are unstable, as they are with just one of time but more than four of space. With only one or two of either, the universes created are too simple to build atoms.

The largest region of the chart contains worlds with two or more of both spatial and temporal dimensions (technically, built on ultrahyperbolic equations), and these again are too unpredictable for life to gain a foothold. That leaves two boxes. The one with three time dimensions and one of space contain only tachyons, particles moving always faster than light. Our 'local island' in this inconceivably vast archipelago is the other stable, sufficiently complex case, with the familiar one dimension of time and three of space.

Tegmark argues that although these other wild universes really exist, they play no interesting part in the cosmos since they must be denuded of anything resembling intelligence, or self-aware substructures (SASs). That's why we are in the kind of universe we find ourselves in—no mind could exist in any other. Do they truly exist? Tegmark chooses to accept the 'principle of fecundity', which declares that 'the physical world is completely mathematical', and that 'everything that exists mathematically exists physically'.

But this argument is not insanely opulent, Tegmark argues, since 'an entire ensemble is often much simpler than one of its members'. Technically, again, the ensemble contains less algorithmic information, the kind needed to specify it in a series of laws or specifications. Perhaps this is just a prejudice of our viewpoint, of course. Tegmark closes his argument with a necessary caution: 'Considering how difficult it is to predict how a mathematical structure will be perceived by a SAS, a systematic study of this issue probably merits more attention than it is currently receiving.'[31]

Meanwhile, more orthodox theoreticians do not cease in their relentless task of creating the universe out of nothing (if only on paper). A July 1998 paper in the prestigious *Physical Review D*, by J. Richard Gott and Li-Xin Li, of the Department of Astrophysical Sciences, Princeton University, suggests economically that the single Universe might be able to create *itself*. This is surely the ultimate in self-bootstrapping. Gott and Li argue that, in general relativity, spacetimes can be curved and connected back on themselves, which permits *closed timelike curves*—paths backward through time. The first instants of inflation might have been a nest of such closed curves, including what amounts to a wormhole from the future universe to its origins. Extraordinarily, they note the possibility that intelligences might create such loops deliberately, which could make our immortal descendants the true creators of a looped but self-consistent cosmos. Here is how they put it:

> *Some specific scenarios (out of many possible ones) for this type of model are described. For example, a metastable vacuum inflates producing an infinite number of (big-bang-type) bubble universes. In many of these, either by natural causes or by action of advanced civilizations, a number of bubbles of metastable vacuum are created at late times by high energy events. These bubbles will usually collapse and form black holes, but occasionally one will tunnel to create an expanding metastable vacuum (a baby universe) on the other side of the black hole's Einstein-Rosen bridge as proposed by Farhi, Guth, and Guven. One of the expanding metastable-vacuum baby universes produced in this way simply turns out to be the original inflating metastable vacuum we began with.*

As a side benefit, this model also provides a natural explanation for the observed arrow of time running only from past to future (despite the closed loops). In a charmingly audacious comment, Gott and Li observe that the laws of physics 'may allow the Universe to be its own mother'.

While we await the closure of the cosmos and the ignition of a deity in the ashes of ruined stars, or even the looping of time and space back to their own birth, perhaps we might use the time—many, many billions of years—to establish our rejuvenated, deathless selves or our machine offspring among those stars in the aeons before their fall. That is, indeed, a prerequisite of projections such as Tipler's and, if those projections are correct, mind will colonise the universe long before the limitations of the spacetime cosmos can be transcended.

seven: life & death transformed

There is a joy, an élan, a spirit to life that is far more important than its duration. That spirit is so closely akin to our health that we might better wish to extend our health span than our life span. Watching those we love sicken and fail, we would wish them longer health over longer life. The enemy is loss and suffering, fear and tragedy. Many of us, therefore, are dedicated to the task of increasing not just a life span, but the quality of life. We wish to add joy and spirit, not just years.

Michael Fossel, PhD, MD, *Reversing Human Aging* (1996)[1]

I certainly don't know whether Carol and I had the ideal amount of overlap, but I would say we approximated it fairly well ... The 'I', the very core, of each person has been incorporated into the other person. Not only lovers, they are now psychologically merged, blurred, blended, and fused, and have come to form one composite entity. This is perhaps why, a few months after Carol had died and I was gazing up at her photo ... I felt myself falling deeply into and through her eyes right into her innermost core, and behind a veil of tears I heard myself sobbing, 'That's me, that's me.'

Douglas R. Hofstadter, *Le Ton beau de Marot* (1998)[2]

Death is intensely personal. Its theft of another human world hurts us deeply if we have shared, even in some small measure, that internalised overlap of which the sublime computer scientist Douglas Hofstadter writes so poignantly. We are more moved by death than by any other passage—except, perhaps, birth. Does thinking about death in the somewhat remote and clinical mode of science distract us from that fundamental agony of loss whose only appropriate answering voice is music and the hard melodies of poetry? I have no wish to evade mortality's personal dimension by speaking of nothing but superoxide dismutases, the mysteries of the quantum, or the resplendent glories of the cosmos.

Still, death in aggregate does not move us quite so fiercely, death on the large scale of those wars and starving throngs we see behind the television screen each night, death as the reaper of millions we do not know and might not like very much if we did know them. Our kindest emotions sag beneath 'compassion fatigue'. And for the young, and honest, death and its mimicry can also provide the fiercest of thrills. Healthy young gangs of men travel great distances to maim each other in soccer game riots, just for the pleasure of it. Vivid computer simulations allow most boys and young men in today's technologically advanced nations, and a few girls, to obliterate imaginary foes in gory detail—even to flash whole cities into nuclear fire. We embrace what terrifies us most. Or we ignore and suppress it. But death will not go away—unless we *make* it go away.

For the first time since single-cell life coalesced on this planet, we are perhaps within reach of doing just that.

So the keen pathos of mortality remains our spur, but we are justified in regarding death, for a time, as an abstract force to be countered with knowledge and determination, rather than appeased by mythic verse and surrender. But solving death, and life, is not a merely technical project. It embraces everything that makes us human. The first immortal generation will not be the children of just science alone, but of law, art, music, writing—all the humane arts.

Although the Human Genome Project will accelerate the knowledge base for direct genetic intervention, there is very much more in an organism (especially a person, as Hofstadter makes us remember, eyes prickling) than is to be found even in a total DNA map. The ethical consequences of the new sciences are formidable, even before the acceleration of science and technology turns into a headlong ascent and death is put to rout (if that ever happens). We need to be prepared well and truly in advance. We need to learn how to think clearly about both death and life.

SCIENCE IN THE SERVICE OF DEATH

Not long ago, a well-informed Australian government genetics research expert told me that 'the Russians are working on ethno-specific viruses'. Imagine a virus tailored to kill only those peoples of the former Soviet Union whose genetic specificity overlaps conveniently with, say, their cultural adherence to Islam. This is precisely what the Russian genetic engineers were said to have been working on. Imagine white racist militias in the USA releasing a deadly virus that locked onto people with those genes which put large amounts of melanin in the skin. Imagine Japanese cultists letting loose not just toxic Sarin gas but a pandemic targeting everyone without the gene for epicanthic folds to their eyelids. The maverick scientist Sir Fred Hoyle has claimed AIDS is just such a military designer virus, escaped or set free as a trial run. (I don't believe that, not for a moment, but in the past Hoyle has been impressively correct—as well as disastrously wrong—with the oddest ideas.)

I doubt that politicians generally will be any more prepared for the news of the impending end of death, and its dangers and opportunities. Genetic engineering, after all, is precisely the discipline that might dissolve our world with some final obscene 'ethnic cleansing' solution. Or eco-terrorists, many of whom have expressed their eagerness to see humanity wiped from suffering Gaia's polluted face, might genetically engineer some sort of runaway omniphage: call it 'green goo'.

THE HU-GOO CATASTROPHE?

Here's the worst and most ironic question of all: would a true immortality treatment, emerging as a spin-off of the HUGO (Human Genome) Project, or telomerase investigations, or some other deep life-sciences research effort, unleash the world's worst plague—an unstoppable tidal wave of endlessly reproducing human beings? It is all very well to mouth wishful platitudes about human intelligence and foresight, but those great gifts have let us down in the past, at least some of the time (although we're still here, despite owning nuclear and bacteriological weapons of mass destruction for over half a century). At other times, coupled to devotion, courage, honour and love, intelligence and foresight have been our salvation and our glory. Tangled through every fibre of human choice, however, are the deep-grained urges of our inheritance as the children of four billion years of driven replicators.

Any one of us can overcome, at least for a time, those inarticulate cravings and imperatives coded into our bodies by survivor genes. Any of us is able to delay parenthood for years or decades past the dizzying adolescence when *Homo sapiens* reproductive programs kick in with their wild hormonal songs, their heartbreaking melodies of loneliness and yearning. Any of us, given sufficient motivation, can choose a life path without any kind of parenting, or indeed any sexual expression at all— there are many voluntarily childless heterosexual people, gays, celibates, paraphiliacs with interesting tastes that preclude reproduction. But not all of us can deny those urgings all the time—and it will be impossibly harder, perhaps, when we literally *have* all of time.

China has managed for two decades to restrict most couples to bearing only one child, within marriage, using an authoritarian and quite brutally invasive policing mechanism. Still, many Chinese citizens break the rules and have more than the one authorised 'little emperor'. (Notoriously, many of the little empresses die in infancy, which does show in a gruesome way that people can deny one of the most powerful instinctual drives even as they pursue the satisfactions of a related cultural pressure.) So what will happen if people learn to live for centuries or millennia, retaining with the aid of science all the fertility and juice of healthy young adulthood?

Here's the worst and most ironic answer of all.

The world would be swollen with hungry human mouths very quickly, even if women only had a new baby or two every 20 or even 50 years. The most prodigious and Promethean technologies might find it impossible to meet our collective appetite for energy, raw materials, living space. Nanotechnology might deliver almost perfectly efficient recycling and cheap power, make the driest deserts bloom, honeycomb the mountains, fill the oceans with self-sufficient artificial cities and the skies with geodesic homelands kilometres in diameter—but sooner or later, even with cheap transport off the planet and into the asteroids, a planet's worth of always-fertile immortals will choke Gaia's life-support systems, or go haywire into lethal carnage.

So, at least, it might seem. Malthus, who long ago forecast overpopulation doom, would prove to be the bleak prophet of utopia.

THE NEW POPULATION BOMB

In July 1998, according to the US Bureau of Census, the global population was estimated at 5,927,383,121 and expected to reach six billion within a year. Using medium-fertility projections, the UN Population Fund estimates that world figures will reach 9.4 billion by 2050 and 10.8 billion by 2150. In fecund Africa, doubling occurs twice as often as the global average, so numbers will swiftly swell from today's 778 million.[3] By 2055, it is hoped, the most struggling nations would attain sufficient economic prosperity that women would adopt 'replacement fertility', stopping at no more than the two surviving children needed to replace mother and father(s).

Yet population momentum will slow the impact of that transition. Two out of five humans in the poorest countries are under 15, readying themselves to add to that huge inertia-driven bulge. Strikingly, while Japan reached replacement fertility as long ago as 1957, its population would take half a century to become stationary (zero population growth) in the early 2000s.

And the world as a whole is nowhere near replacement fertility: the current level is four surviving children for every couple rather than two. Not even the appalling epidemic of AIDS will scythe that swelling human harvest. 'Even with the projected losses,' George Moffett has noted of the AIDS plague, 'population growth rates will remain high and population doubling times will be increased by only a few years.'[4] On the other hand, if heterosexual people most at risk of AIDS in the Third World adopt condoms as a protective measure against disease, that might in turn reduce runaway overpopulation as a side effect.

Conservative UN projections see the current 560 million people older than 60 soaring even without novel life-extension techniques. Today, with about a tenth of all people over 60, will seem youthful in retrospect by contrast with 2150's population, for 31 per cent are expected to be in that category—even *without* any drastic life extension. So what happens if science does extend life span significantly? Even replacement fertility would be excessive for a groaning planet, since those ageing adults whom the new babies were meant to replace will still be alive and kicking, consuming, polluting with their energy demands, and perhaps making even more babies. Prophets of population doom like Paul and Anne Ehrlich might be proved right after all.

It is difficult, though, to assess the population consequence of radically altering our life span, since much hangs on whether the very, very old would still be fertile, and if so whether they would wish to indulge their ability to produce more babies. Suppose fertility is retained. A lineage of deathless couples bearing just two children every 25 years (that is, in each traditional generation interval) creates the same population growth profile as today's Third World mortal parents who have four children before dying. Despite the inevitable deaths that cull mortal numbers, both mortals and immortals quickly start doubling with each generation—geometric or exponential growth. Both paths, of course, constitute a population explosion, and in a finite, bounded world (even one with cheap space travel and abundant resources in the asteroid belt) will need eventually to be modified by curtailing the number of children born.

Swedish transhumanist and neuroscientist Anders Sandberg offers these further reflections:

If people just stopped dying and having children, we would simply see the current age curve move upwards with one year per year; if people continued to have children and retained a slight risk of dying we would get a cluster of people born in the twentieth century growing ever older and slowly dwindling, and a long tail of children born after that growing up. In the really long run you would have a kind of exponential distribution with a very long tail and low numbers of people at a certain age ('I'm the only person born in the 35th century I know'). [5]

THE POPULATION PROSPERITY ENGINE

Is it true, though, that the new demographic transition into extended or rejuvenated life would impose terrible consequences upon those alive in such a brave new world? Techno-optimists such as the late Julian Simon, a professor of business administration and doyen of a libertarian think tank, the Cato Institute, argued vehemently that economic growth in human societies booms alongside population increase, without leading inevitably to resource depletion and environmental degradation. How so? Because we are cumulatively intelligent creatures, whose cultures learn new and more efficient ways to shape the materials of the world to our needs. Once we were obliged to accept only what came our way from nature's bounty; then we learned to draw abundance from the land; now we build altogether new materials. Every year we work smarter and cheaper, with wealth increasing for all (for some much more than for others, it's true) as market signals enable us to maximise our productive powers. [6]

The economic tragedy of ageing and death is that so much human effort and resource is expended in education and training that's inevitably lost within a couple of decades. Genetic memory alone is automatically transmitted more or less unblemished down the centuries, the millennia. Cultural memory is compressed, shared, committed to writing and other external storage media—but its fine-grain detail perishes, in large measure, with every human who falls. We pour increasingly greater quantities of treasure into teaching first the fundamentals and then the sophisticated expertise of a thousand trades and professions and, in each case, within 50 years that treasure is wasted in death.

Suppose it need not be that way? Suppose the accumulated knowledge and, better still, the hard-won wisdom of the years were retained—not in written summary, not in tales passed down from parent or tutor to child, but in the canny brain and sinews of each vastly long-lived adult?

THE PRICE OF LIFE

Since we still do not know if extreme longevity through technology is feasible—let alone rejuvenation for those of us already on the downward slide into mortality—it is difficult to offer a realistic estimate of its costs. Almost certainly, more will be required than a simple inoculation, even if that included plasmids or retroviruses able to rewrite 'death genes' or rejig telomerase production. At the extreme, the next hundred years or more will become dominated by an immensely expensive new life-extension industry.

That course has been sketched in a number of thoughtful and vivid futuristic novels. In Joe Haldeman's *The Long Habit of Living* (1989), a well-written techno-thriller, the Stileman Foundation holds proprietary command of a rejuvenative technique available only to the rich. For an upfront fee of a million pounds sterling, plus all the assets of the prospective patient, successful or lucky 'ephemerals' can buy an extra decade of vigorous life. Divested of their fortune, they must strive anew each time to rebuild their stake for the next round. Haldeman's story inevitably probes the system's points of weakness for corruption and abuse, but makes the useful point that either private or state interests might drastically limit the power the endlessly young could accrue. Perhaps no less usefully, Haldeman hints at the excruciating agonies that could be entailed in rebuilding aged bodies.

A remarkable Tolstoyan trilogy on the theme of the settlement of Mars in the next couple of centuries, Kim Stanley Robinson's huge *Red Mars*, *Green Mars* and *Blue Mars* (1992–6), is populated by anti-agathic pioneers who grow ever older and ever stranger, but usually wiser. Perhaps the most convincing depiction yet of advanced longevity medical practice and its consequences is Bruce Sterling's 1996 novel *Holy Fire*, set at the close of a swiftly changing twenty-first century. With William Gibson, Sterling virtually created the cyberpunk genre of post-modern science fiction. His comments are astute and far from complacent:

A serious life-extension upgrade was a personal crisis to rank with puberty, building a mansion, or joining the army.

> *The medical-industrial complex dominated the planet's economy. Biomedicine had the highest investment rates and the highest rates of technical innovation of any industry in the world ... The scope of gerontological research alone was bigger than agriculture.*
> *The prize was survival.*[7]

And the price of that prize? Ceaseless trials and partial successes. Sterling is grimly funny:

> *Medical upgrades were always improving, never steadily, but with convulsive organic jumps. Any blue-chip upgrade licensed in the 2090s would be ... about twice as effective as the best available in the 2080s ... As for the 2050s, the stunts they'd been calling 'medicine' back then (which had seemed tremendously impressive at the time) scarcely qualified as life extension at all, by modern standards.*[8]

Mia, Sterling's ancient, quite wealthy and *very* careful oldster, tries a new experimental method that returns her youth at the cost of obliterating her memory. 'Neo-Telomeric Dissipative Cellular Detoxification' is a horrendous multi-stage operation. Mia's gut is puttied closed and her lungs drowned in oxygenated silicone, slaying most of her inner flora. Deeply unconscious, her brain is scrubbed with sterilising fluid. In a tank of gelatinous gunk, her body is returned to the dependency of foetal life so that bacteria, viruses and prions can be hunted down and killed. DNA repair is started at last:

> *Intercellular repair required a radical loosening of the intracellular bonds, so as to facilitate medical access through the cell surfaces of the corpus as a whole. The skinless body would partially melt into the permeating substance of the supporting gel.*[9]

Vastly swollen, this mindless sausage meat is pierced by access tubes. All manner of specific attacks are launched in various organs and tissues. Pre-cancerous cells are located and purged. Finally telomeres in every tissue are extended to youthful length, 'tricking the ageing cells into believing the fiction of their own youth'.[10] Mia undergoes this horrific procedure but, in effect, loses her self. It is a memorable opening to Sterling's searching look at a future where age-extension and even immortality can be had for a price, in this case extreme compliance to convention, self-sacrifice and 'good citizenship'.[11]

The Dead Hand of the Past

Is another price of immortality, if memory can be retained during rejuvenation, that it must be a recipe for crushing stagnation? As things are today, technical knowledge doubles even faster than human numbers. The fresh young minds of children are specially structured to take in tremendous amounts of raw and processed information, to learn new words for new ways of looking at the world. An entire generation of researchers struggles for decades to create an impossibly difficult innovative way of perceiving physics or chemistry or economics, and a pristine student intake at college effortlessly gulps down that shocking paradigm shift and moves forward creatively, within what seems to them the most natural perspective.

Science, no less than the arts, advances as the old bulls weaken and retire. So what happens when the old at the helm no longer weaken, when nobody is obliged to give up tenure or office due to physical exhaustion and slow mental decline? Is this not a return to the grim days of repressive priesthoods and principalities, when opinions are set in concrete and no disruptive new angle on the world is permitted to ruffle the timeless surface of The Way Things Are Meant To Be?

Certainly, that is a risk. It is not, though, I think, a very persuasive picture. Julian Simon's potent market forces would surely step in to topple a rigid gerontocracy, as they are doing now in China. If the old refuse to learn, they surrender possession of the finest novelties to their younger rivals. Knowledge is indeed power, and new knowledge brings surprising access to new power.

Besides, why must an ageless, rejuvenated gerontocracy be mentally and emotionally rigid in this way? It is too early to know, but we might be taking a false logical step in assuming that the old in a deathless world must automatically resemble today's crusty elderly. One reason older people today tend to cling to views acquired in their own youth and early maturity is that the brain literally gives up its plasticity. The neural wiring is trimmed from the earliest months and years like a manicured topiary bush, shaped by each child's environment to create an adult well adapted to a certain historically contingent time and place. Eventually that specialisation—with its concomitant loss of flexibility—hardens into place as the ageing brain suffers damage. Neurons tangle in plaque, dying and not replaced. Yet this physical deterioration is precisely what an immortal generation will need to halt in order to forestall death. Telomerase therapy, even nanotechnological machines like hordes of clever benign viruses, will maintain the repair and health of neurons and their vast network of knowledge—indeed, of self-awareness. If mental rigidity is a risk for the deathless, it will be a psychological rather than physiological hazard, the kind that already afflicts 'young fogies' who choose to squeeze their eyes shut in the face of novelty.

Hope I Die Before I Get Old

For all that, the political consequences of massively extended youth might not be comfortable. On the one hand, a world in which everyone retained the piss-and-vinegar zeal of adolescence, in which male 'testosterone poisoning' was *never* quenched in the hormonal ebbing of late adulthood, might be an alarming place to live in. I suspect this risk can be overstated. Even diehard criminals in their late forties or fifties are physically strong enough and brutal enough to maim and kill their foes, but it seems that they often mellow with experience. Time simply teaches us lessons of prudence that the innocence and ignorance of youth cannot know. Happily, we also learn 'people skills', learn to ignore the foolish herd-mentality of the young, learn courage in taking chances that once might have frozen us solid in terror of imaginary consequences. To retain those skills in a body unhurt by the years—ah, 'Bliss was it in that dawn to be alive,' as Wordsworth did not quite say, 'but to be old was very heaven!'

On the other hand, the old are full of stored spite and cunning, unwilling to yield. NASA scientist and award-winning science fiction writer Geoffrey A. Landis has commented, 'The more I think about it, the more I believe that extended life would actually be a horrible curse to society.'

If people lived for 200 years, I think it is inevitable that the people who hold all the power would be the 180-year-olds.

The drive to hold on to power is just too strong. The older people get, the craftier they get in their ability to hold power, as well as the more conservative, set in their ways, and resistant to change. Basically, the older you get, the more energy you have spent in adapting to the world the way it is, and thus the more change will be uncomfortable because it upsets all the adaptations you've made.

The penalty to society would be worse, I think, than merely proportional to the increase in life span, since in a society where people live to 200, when the 200-year-olds finally do relinquish power, they will be replaced by 170- and 180-year-olds, who are themselves already calcified in their thinking.[12]

A more hopeful reading is a charming short story by the late George Turner, 'Not in Front of the Children' (written when he was over 70), which proposed a world of self-segregated *regna* or generations sustained by costly anti-geriatric treatments that merely postpone the ill effects of ageing.[13] One regnum is the Intellectual Women, followed by the Homeloving Mouse regnum, the neo-Victorians and the Liberated. 'A family might have eight generations extant, all loathing each other, and not cordially. In particular, the young detested the old whose signs of age were flaunted on their faces, and refused to recognise kinship.' Luckily, matters improve, at least for Turner's adolescent heroine.

What, though, if those signs of age are *not* flaunted, are not easily determined by quick inspection? Another science fiction writer, Larry Niven, suggests that the very old *will* be identifiable—paradoxically, by the extreme grace and economy of what we today would regard as their 'youthful' motion and poise. After hundreds of years of ease in a healthy body and a world of reliable objects, the ageless old will not bump into furniture or waste effort in nervous jittering. It is a compelling and uncanny image of the immortal condition.

INSURING FOREVER

Several very practical questions are usually raised at this stage in any enquiry into the anti-agathic condition. Some of them go away quietly when you look at them hard. Others defy us, since it is impossible to answer them in our present ignorance of future possibilities. Still others fall somewhere between. We can offer answers, but without a great deal of confidence in them.

Who will pay for the pensions, if everyone lives for hundreds, let alone thousands, of years? It's bad enough now, with the Baby-Boomers greying and the succeeding generation fewer in number than their parents, who are nearing retirement age. Old-age pension schemes handled by governments, and superannuation or life assurance plans run either on behalf of companies or individual investors, largely depend on certain age-old verities remaining true. Actuaries have worked out the grim statistics of expected death rates among clients, and on that basis they juggle payments, rates of return earned from investing that income, and sums paid out over varied intervals.

Insurance policies are many and varied, from whole-life and term to endowment, plus blends of these. Fixed-term and endowment policies would be untouched by drastic life-extension, but whole-life, by definition, would alter if longevity were changed. Premiums set under one set of expectations would no longer be appropriate. On the one hand, how many people would continue paying premiums (or start in the first place) if they had a virtually unlimited life expectancy? On the other hand, certain annuities written prior to the breakthroughs and which become effective from, say, age 65 and continue until death, would be obliged to pay out regular sums, perhaps tied to a cost-of-living index, forever.

Of course, that is unrealistic. Enabling laws would be changed to protect corporations stuck with policies underwritten, as it will suddenly appear, more adversely than any in history. The impact will be powerful, since for at least the last 30 years the value of life policies in the developed world has exceeded national incomes. By 1990 in the USA, for example, according to *Encyclopaedia Britannica*, *nearly $9.4 trillion of life insurance was in force. The assets of the more than 2,200 U.S. life insurance companies totaled nearly $1.4 trillion, making life insurance one of the largest savings institutions in the United States. Much the same is true of other wealthy countries, in which life insurance has become a major channel of saving and investment, with important consequences for the national economy.*[14]

In the United Kingdom, the life insurance industry and pension funds hold (and invest) much of the nation's wealth. The latest available UK National Statistics (1996) show these remarkable figures: in a year when gross domestic product was 601.7 billion pounds, and consumer expenditure was 376.7 billion, pension funds held 564.1 billion pounds, long-term insurance funds held 554.2 billion, and other non-long-term funds accounted for 77 billion.

That is still not true in the Third World, despite massive efforts to launch insurance as part of the way of life in those countries. In many, the mere fact of buying a policy is regarded as fatally bad luck, an explicit invitation to ill fortune. For the world's big life insurers, the coming of an effective cure for age will be *karma* of a kind perhaps still incalculable, for good or ill. And the impact on varieties of investment is also problematic. If long-lived healthy people choose not to save in this traditional fashion, perhaps instead they will pour their accumulating wealth into growth stocks and ventures. Or perhaps capital will be frittered away on luxuries and direct consumption. Or, as we shall see, the very terms of this discussion might be rendered moot by radical shifts in the technological basis for making and distributing goods.

This is not as theoretical as conservative thinkers might wish to imagine. During the last few decades, actuarial curves have slowly shifted as more and more people remain alive for a larger and larger proportion of the maximum human life span. This change is putting pressure on both governments and corporations. Imagine what happens if people's lives *double* in length! Corporations will crash. National economies will be in crisis. Who will pay?

The question is based, in some measure, on fallacious premises. At the moment, matters are indeed getting out of hand. That is because extended life now goes hand in hand with increasing debility and risk of long-drawn-out expensive illness. The aged are increasingly doomed to a prospect of perhaps two or even three decades of discomfort, failing abilities, weakened senses, eroded minds. Often they end lonely, deserted, in a state of living death hooked up to costly high-tech life-support systems. This surely *is* the curse of Tithonus. But what we have discussed in this book is almost certainly the technological path *away* from such a distressing close to life.

Granted, perhaps none of the medical and other solutions to age and death that we have considered will come to pass. Perhaps obstacles as yet unseen will arise, preventing anti-agathics from having the rejuvenative and prolongevist effects I have predicted. But if such methods *do* work ... everything is changed. And mostly for the better.

Who will pay for those who retire automatically after 40 or 45 years in the work force? Why, nobody—because people with extended life spans will no longer be obliged (or entitled) to retire. Why should they be? At the moment, the elderly do so because age 65 or 67 (or 60, or 70: it varies in different countries) is the somewhat arbitrary point at which workers *by and large* are considered no longer cost-effective by management. Nor is this sheer bloody-mindedness. The expenses of raising a family have been met. Often, a family home has been purchased (and will sometimes now be exchanged for a smaller unit).

In many jobs and professions, the physical or mental strength and agility of youth is at an end, and people simply cannot do the same jobs any longer. In others, skills learned 40 years ago are so obsolete that not only do they no longer pay their way, they actively obstruct innovation. Of course, mental retooling is increasingly required in many professions at younger and younger ages, so that some people find themselves on the scrap heap at 50, 40 or even earlier.

Once ageing is beaten, some of this quickly changes. If sheer physical strength is called for, it will still be there. Mental prowess will no longer fail routinely and inevitably. Yes, retraining will be required for many jobs, but the healthy old will be at no special disadvantage over the young. Habit might make teaching old dogs new tricks rather harder than just starting over with a batch of new pups, but steadiness and experience of life and the world's quirks might compensate. If anything, the young will chafe under the tenure of their wily, unbudgeable elders. These changes will arise slowly enough, I suspect, that cultural mores will adapt. Yes, there will be difficulties. It won't be easy during the transition. But if the alternative is to grow feeble and then die—which choice will *you* make?

Another frequent question: where will all the new jobs come from? If the old remain alive for centuries, surely the pressure on employment opportunities will be ferocious and malign! Again, this is an odd fear. Jobs are just what humans must do with their brains and hands in order to create the goods they consume, the services they need to live pleasant, comfortable lives, the obligations they share in keeping the environment clean, healthy and sustainable. Once there were only half a million humans on the whole planet, and none of them lived as well as billions do today. True, many other hundreds of millions now live in shameful squalor and want, but mostly that is due to inequities of distribution in the goods we have readily available or could easily produce. If billions more remain alive and healthy, contributing to the economy of the globe, they will pay for themselves just as their children will— by working, by helping each other.

You might have noticed another hidden taken-for-granted premise in that argument. Will the need for work as we know it persist? Almost certainly not. After all, we are positing a high-tech future capable of the medical and nanotechnological miracles needed for a prolongevist life. The implications of scientific change for the work force go far beyond the space available to us in this book (I have sketched them in *The Spike*). Suffice to say that several lines of swiftly accelerating change in computation, genomics and molecular manufacture make it likely that by 2050 very few jobs will be unchanged. Most of today's work will be either gone or on the verge of vanishing.

How is that possible? Briefly: many experts expect affordable computers with the processing power of a human brain by 2030 at the latest. If our grasp of the mind and AI are sufficiently advanced by then, those machines will be the wedge into an awesome productive capacity that will simply bypass human labour. This seems an atrocious vista to those currently out of work or about to be, but that is because we now see everything through the experience of age-old scarcity economies. Once 'smart' machines take over the world's work, most of the basic requisites of living will become absurdly cheap. Indeed, once the first programmable nano-assemblers or matter compilers are built, these astonishing machines will be able to replicate themselves at very little cost. Thereafter, many of the traditional causes of conflict and strife will evaporate, for everyone will have material comfort at little or no expense, and while there will be few jobs in manufacturing or even information technology (because those jobs will be fully automated, as was predicted prematurely in the 1950s) the goods will continue to roll out—or be compiled at home.

So we will need to reorganise the economic bases of society, and nobody should expect that to be easy or painless—but neither will the ageless face the problems of today writ large. Problems, perhaps, but new ones. Constructing utopia (and preventing horrific local warfare between restless new tribes and gangs) will give them something to do with all their new free time ...

Does this mean money will vanish? No. Some resources will always remain limited. Choice land will be one. If anything, it seems likely that financial systems will become even more pervasive and embedded than they are today, for many flows of goods and services might be monitored to a degree we cannot yet comprehend. People might have to learn to live in what has been called *the transparent society*.[15] Alternatively, cryptography and a vastly overgrown Internet might permit all manner of odd activities to proceed undetectably by the forces of law and order and crime alike. The world of the immortals will be increasingly strange. Luckily, our minds will probably recover the agility of youth. Even so, many people will bend under the shock of relentless change. And many of those will simply choose, either in despair or with dignity, to remove themselves from the project of endless life. That will be their privilege. But they will be missing out on some very interesting times.

WILD POSSIBILITIES

A very modest proposal for the first immortal generation, and its successors, is that world population will first pass through a demographic transition into replacement fertility, plateau out within some 50 years, while adding an additional doubling as the old fail to die. Will rejuvenated, healthy women add their own new generations of babies to the crush? The most conservative expectation is that they will not, because fertility is limited by the number of follicles retained in the ovaries. These have existed since a women was a foetus, and are not renewed. It is possible that telomerase therapy might reawaken the stem cells in expired ovaries, but that seems more like a prescription for cancer. It is also possible that nanotechnology could rewire an aged body to a renewed fertility, but that is an option most people might care to disregard, or invoke only infrequently.

But we need to remember that these high-tech innovations will arrive bundled with many others. At the moment, human cloning (in the sense of growing a new baby from an adult's differentiated tissue) is still widely regarded as repugnant and even illegal. In March 1997, just ten days after Dolly the cloned sheep entered the headlines, President Clinton uttered a predictably pious executive announcement in the Oval Office. 'Each human life is unique, born of a miracle that reaches beyond laboratory science. I believe we must respect this profound gift and resist the temptation to replicate ourselves.' That opinion resonated within many hearts, but I do not think it will long remain ironclad. Somewhere, soon, human babies will be born from this technology, carried in human wombs, and it will be quite apparent that they are just like all other babies.

A little later, artificial wombs will sustain the lives of babies for the full 268 days, on average from conception (cloned or otherwise). At first they will be used to keep alive children who otherwise would have perished in the womb—but sooner or later this technology will surely become commercially available to parents who do not wish to bear their own children, or cannot, or are too old to do so. And after that—why, even if rejuvenated women have lost their own follicles, and it remains too dangerous to try to regrow them, there will always be the option of having cloned or recombined children who spend the first months of their lives in a life support system as snug as any natural womb.

If that sounds far-fetched, it is tame compared with other possibilities that science holds out to us. If it ever becomes feasible to replace parts of the failing brain with cyber-implants, designer neural packages that take up the burden of failing brain cells as artificial hearts and kidneys are beginning to do with those organs, we may reach a stage where people are as much prosthesis as original. If nanotechnological computers as small as matchboxes, or even tinier, can contain as many working neurodes as the brain contains neurons, we might all carry a kind of backup redundant copy of each module in our brain's circuitry. What happens then when the protein brain fails in senescence, dies in some last hopeless contest with infection, or is smashed in an accident? If the backup 'black box' system is multiply wired throughout the organic brain, you won't even know that your original brain has perished!

This might sound utterly incredible, but nanomedicine pioneer Robert A. Freitas puts some numbers into the speculation and makes it sound quite reasonable—given the assembler technology of molecular nanotechnology:

> Consider that a nanostructured data storage device measuring $\sim 8,000$ micron^3, a cubic volume about the size of a single human liver cell and smaller than a typical neuron, could store an amount of information equivalent to the entire Library of Congress. If implanted somewhere in the human brain, together with the appropriate interface mechanisms, such a device could allow extremely rapid access to this information. A single nanocomputer CPU [central processing unit], also having the volume of just one tiny human cell, could compute at the rate of 10 teraflops (10^{13} floating-point operations per second), approximately equalling (by many estimates) the computational output of the entire human brain. Such a nanocomputer might produce only about 0.001 watt of waste heat, as compared to the ~ 25 watts of waste heat for the biological brain in which the nanocomputer might be embedded.[16]

Indeed, having a constant on-line backup unit inside your head (or beaming its vast number of experience-processing bytes to a distant safe site via a kind of cell-phone grid, perhaps linked worldwide through satellites and optic cables) will allow you to shift your point of view, of consciousness, between your body and anywhere else you choose. You would become a disseminated mentality, your awareness stretched effortlessly across the entire globe (but you will need to get used to the psychic whiplash and 'motion sickness' of changing point of view too quickly or too far). Once your mental functions are at least aptly coded into a machine, it will also become possible to create coding 'handshaking' protocols with the machine substrates supporting other people's extended consciousness. It will be a kind of artificial telepathy, an extended cyber-empathy in which you might merge at least the fringes of your self into that of a loved friend—or many of them. Ultimately, we might find that enhanced humans have become the benign and transcendental equivalent of *Star Trek*'s Borg, a sort of colony organism with multiple minds, as firewalled or as fluent and intermingled as they wish to be.

WILDER POSSIBILITIES

If such apparently extravagant changes enter the human shopping menu, matters will not stop there. If you can upload or back yourself up in a powerful computer, you can also clone yourself mentally. That is, you will be able to purchase a machine (or space in one) and 'xerox' yourself as an independent copy (call it 'xoxing', a term coined by a witty transhumanist who must have had a fondness for Dr Seuss). The xox won't be you, exactly, because the history of his or her consciousness would start to diverge from yours the moment the copy was complete. But the two of you would be closer than twins.

Is this an enticing possibility, or a hideous one? For many, the shudder they feel at contemplating this choice is sufficient answer. Others would wish for nothing finer than the creation of a duplicate self, or a whole family of them. My point is simply this: there are enough people who would happily take up this option, if they can afford it, that the whole question of overpopulation bursts up anew. Forget telomerase therapy, and piecemeal fixes to a body smashed remorselessly by free radicals and virus attack. If uploaded immortality and xoxing are on the agenda before the end of the twenty-first century—as some computer scientists such as Hans Moravec believe—the first immortal generation will gradually morph into beings we can scarcely imagine today. And if we can just hang onto old age long enough by using those intermediate technologies of repair and life extension, we might be among their number.

THE FUTURE IS ANOTHER COUNTRY

So do we need to worry about the consequences of biological life extension? Perhaps only in the short term. Tim Freeman offers the following logical if somewhat chilling projection, conceiving of a future world where technology has opened fresh realms in outer space to both humans and artificial intelligences. Freeman, who holds a PhD in computer science from Carnegie Mellon University and is involved with the cryonics company Alcor, compiled the Internet FAQ (Frequently Asked Questions postings) on the topic of cryonics, preservation of the dead at very low temperatures in hopes of resuscitation and treatment in the future:

> Ultimately, the same forces will have to control the population of sentient entities near the Earth that control the populations of animals now. If you don't have the resources to support kids, and you have them anyway, they'll starve or you'll starve. Life expands to fill the available niches.
> However it's pretty unlikely that those sentient entities will resemble humans much in a century or two. The process of generating 'kids' is unlikely to resemble what we are accustomed to. There is foreseeable technology that can build things that are orders of magnitude better information-processors than a human brain, and market forces will cause that technology to be used. The choices for people like us will be either to change radically or to become irrelevant or dead. I haven't heard any scenarios plausible to me that would leave any biological life on this planet in two centuries.[17]

This is not the Doomsday scenario it might sound. Rather, it is a clear-eyed, unsentimental acknowledgment that the future will be different—*drastically* different. Does that mean it is not the kind of place you or I would wish to spend time, let alone backed-up and endlessly upgraded eternity? Perhaps—but we will have the opportunity to grow and change as innovations tumble from the cornucopia of increasingly intelligent technology.

Nor does this imply any loss of humanity, in the deepest sense of that loaded word. Consider the following bracing (or, as some will see it, excessive) vision offered as long ago as 1984, in *Cryonics* magazine, by Hugh Hixon, whose actuarial calculations show that without normal ageing we might live for many thousands of years: 'I plan on having my identity duplicated (or shared) among a large number of curious people and other constructs, a versatile manufacturing complex, a couple of armored divisions, and a space fleet, to be used as necessary to create co-operative environment, all units to have shared goals and considerable autonomy to implement them.'

Armoured divisions? Space fleet? What kind of nerd's wet dream is this? That might be, at least, one's first reaction, but I think it is unfair and unkind. A member of Alcor's Board of Directors since 1982, and a retired USAF Captain with degrees in chemistry and biochemistry, Hixon is simply carrying forward into a plausible if extreme future the basic tenets of game-theory and ecology. His vision is not one of imperial war-mongering amid the stars. Far from it. As he notes (rather sweetly, I think): 'I'm going to make friends.'

ARS MORIENDI, ARS VIVENDI

Hundreds of years ago, western Christian cultures developed a sophisticated *Ars moriendi*, or art of dying well. For now, we still need those arts. Thanatology has grown as a theory and practice in the last few decades, as death changed from abrupt commonplace disaster afflicting the young to a long drawn-out ebbing of life in the ancient. Experts in dying and grief chart the psychological response to death's approach: denial, followed by resentful rage against the living, followed by bargaining, then sunken depression, and at last, if the dying person is fortunate, a stage of peaceful acceptance. The new technologies, or their promise, are going to disrupt that hard-won apparatus of comfort.

The cautious and the sceptical might regard the entire tenor of this book as evidence of arrest at the stage of denial. Death is so terrible, they will say, so final, so unappeasable, that naive science groupies flee from its implacability into a fantasy of technical redemption or reprieve. There is doubtless some truth in this charge. As yet, no method recommended by science (or by magic, affirmation or prayer, for that matter) has managed to extend human life beyond the naturally evolved limit of about 120 years. But something has changed lately: laboratory scientists *have* extended the life expectancy of at least some living creatures, and by a striking amount. And normal, healthy human cells have been literally immortalised in culture. This is not the abolition of death—but it looks very much like the first steps toward that goal.

Not that death will ever be abolished, of course, not literally. In the context of our discussion in this book, you will recall, terms like 'deathless' and 'immortal' have been given a clearly limited, defined meaning: lacking a pre programmed life span, not *totally immune to destruction*. An immortalised HeLa cancer cell culture in an oncology lab can be killed by denying it sustenance, by disrupting it with radiation—or simply by mashing it brutally with a hammer. On top of that obvious proviso— that people whose ageing clocks have been turned off or reversed will not die routinely just of 'old age'—we have to remember that identity itself is in flux with the passage of years, let alone centuries.

Are you the same person you were as an infant? In some respects, obviously. Even a victim of massive retrograde amnesia, forgetting everything except the rudiments of language, is in some sense the same person, now drastically impaired and reduced. More of character and personality, we are slowly learning, is 'in the genes' than was imagined halfway through the twentieth century. So monozygotic twins really are more often similar in temperament, interests and tastes than one would expect by chance, even if they've never met.

On the other hand, we surely diverge from that basic genetic template as we gather rich experiences. Even if you retained all your memories, would you truly be the same person in two centuries' time that you are now? In another thousand years? In 10,000? If you managed to survive until the death of the Sun in five billion years' time, rewriting parts of your DNA code or shifting into silicon or quark soup inside a handy neutron star, compressing and selecting your colossal memories, splitting and remerging your multiple agents and partial selves, how much of you will be able to say: 'Yes, I am the same person who began life near the cusp of what we then called the twenty-first century, yes, here I am still, deathless and essentially unchanged'?

Still, these admissions of the limits of identity and the inevitable impact of cumulative change are a far cry from embracing death as the right and proper end to a short human life. Right now, of course, we are in a psychologically very oppressive moment. The wisest and most sensitive hearts and minds of all human cultures have struggled with the shocking fact of inevitable death. Facing up to our demise is, in a way, the enduring project of humankind. It has never been avoidable, and even long life and robust ageing offer no final consolation for its severance. Little wonder that, far from accepting the current finality of death, most of humankind is quite literally mired in denial. While few dispute the brute fact of death and decay, which nobody except the insane or very childish can deny, many do refuse point-blank to accept the utter termination of consciousness which our corporeal decay ordains.

It is especially ironic, after all, that the most vehement opponents of scientific research into extended longevity—into 'immortality', in the restricted sense we have adopted—should be precisely those who foster belief in some other kind of augmented life, beyond death's door, a mysterious deeper life that allegedly surpasses the fact of physical corruption. What is religion's contribution to the debate, in the end, but a systematic and subtle blend of anguish, terror, bargaining, hope, and capitulation and embrace of the inevitable by denying its reality?

I am in no position to judge the validity of these claims, nor is any mortal, for none of us has returned from the grave to be probed on television or in the laboratory (although the believers in reincarnation would dispute even this). Still, I do not see in the avowals of believers any fundamental objection to the enhancement of life's prospects in this world. Perhaps it is true that mortal life, the only kind we know, is just a preliminary for fuller life in a more elevated condition (or perhaps for endless torment if our choices during life were evil). Well then, the longer we have to savour the rich joys and tests of earthly life, the more opportunities we gain to mature, love, help, build, to take and share responsibility.

Today we weep to see some child doomed to piteous early death by leukaemia. Adults rush to contribute bone tissues that might reverse the errors in the child's genes. In another century, we will feel no less anguish at the sight of ageing and death in anyone at all. Life is as replete and meaningful as we make it, and if we can share in the glories of the world for a thousand years instead of a mere 70 or 80, I cannot imagine that a loving deity would resent our tenure here, or punish us for living well and long.

Professor Arthur E. Imhof, a historian of the Free University of Berlin, has lately reflected on the need for a new *Ars bene moriendi*, one suitable for a time when already we have effectively doubled the traditional expectation of life. Imhof is by no means a technological prolongevist. If the promise of indefinitely extended life does, alas, prove to elude science, we will be well advised to embrace his humanist prescription. 'Let us transform *every* one of the years gained into fulfilled ones, taking advantage of our immense technical, economic, and cultural resources, and then let us die a natural death.' But if it turns out that we need not die after all, if extended life is gifted us in technical solutions to ageing undreamed of by ancient alchemists and mystics, then Imhof's further advice still seems to me sound and bracing, an *Ars vivendi* for an all-but-immortal future population.

> *Gained years are not necessarily fulfilled years. We have to fill them with meaning. If people spend their lives interested only in physical activity but not in spiritual-cultural matters, they should not be surprised to find themselves confronting a great spiritual emptiness when their physical powers wane in old age and they do not know how to fill the extra days, months, and years. But this need not be so: the situation could be prevented with a lifelong cultivation of spiritual and artistic interests.*[18]

Those interests—obsessions, perhaps, indulged across centuries of freshly garnered information and insight—will include, for many, the topics we have touched on in this book: the nature of mind and consciousness itself, of life natural and artificial, and of cosmos great and small, the quest for intellectual clarity, and perhaps our emerging role as custodians and shapers of the whole future universe. And within that passionate quest for understanding, we will surely also seek to nourish the quiet, serene truths Imhof recommends: arts of living well, and if death is, after all, finally unavoidable, however long postponed, of dying well.

WAITING FOR THE BREAKTHROUGH

Meanwhile, what are we to do? Death and ageing are not yet defeated, nor is the prospect immediately in sight. Some hopeful commentators such as Michael Fossel claim that effective telomere therapy will be available 'between 2005 and 2015'[19]. Others decry this as 'feverishly optimistic'. I have no opinion, not being qualified to offer one, although I find it notable that experts in the relevant fields do not dismiss expectations of extended longevity. V. P. Skulachev, of the Belozersky Institute of Physico-Chemical Biology, Moscow State University, concluded an overview article in a special 1998 issue on *Telomere, Telomerase, Cancer, and Aging* of the authoritative journal *Biokhimiya (Biochemistry)*:

> *Characteristically, a culture of immortal cells can be produced only by breaking at least three genetically different mechanisms. The last of these mechanisms is that of switching off telomerase. The actual number of such barriers in the body can be even greater. However, the mere fact that their number should be finite can make those seeking human immortality optimistic.*[20]

Suppose we conclude the optimists are justified in fixing a near-future date for the first medical availability of prolongevist treatments. Would people try them? Would anti-agathic treatments be shunned as the work of the devil, or sought out as eagerly as fancy skin creams that already cost an arm and a leg but don't do much except smell wonderful? We have a clue in the wildly excited response to Viagra, the heart pressure drug that in 1998 serendipitously proved to offer two-hour erections to some 50 or 60 per cent of those luckless men crippled until now by sexual impotence. Legally available at first only in some countries, the drug was boosted by massive media coverage, sought out by millions of middle-aged and elderly men and their partners, smuggled into countries where it was not yet freely obtainable, sold for up to $1000 a pack. Clearly, when it comes to recovering the powers and stamina of youth, many men stampede with money in their hands. Would it be any different with a treatment that genuinely slowed, halted, or even reversed normal ageing? I don't think so.

Even a drug that successfully returned luxuriant hair to the heads of the bald would transform the world in a trice. (In fact, in January 1998, Professor Angela Christiano of Columbia-Presbyterian Medical Center in New York announced a chromosome 8 gene, named *hairless*, in the journal *Science*. *Hairless* is a transcription factor, influencing other genes, and its isolation might lead within five years to a very lucrative cure for male pattern baldness.) Against the lab-tested promise of endless life, not to mention baldness, I do not believe religious or other qualms would long stand in the way. And even if there are portions of the community who hold out against such treatments, they will be literally a dying breed.

Perhaps militants might go so far as to bomb longevity clinics, as some extremists now murder abortion medicos and their patients, but that phase would soon pass. Within a century or two, after all, the long-lived would simply comprise a large, wealthy and influential proportion of any community where the treatment is available. Those who object on moral or aesthetic grounds to extending their lives in this way would surely be permitted to decline it—although there might be some intriguing legal issues raised if doctors withhold life-extension measures even at the patient's request—but their numbers will fall with the years. Life on earth, for humans (and perhaps chimpanzees, gorillas and our favourite animal companions), will be an endless golden afternoon ... or perhaps, better still, a fresh morning with endless opportunities.

Hanging on for Grim Life

And in the meantime? How tragic to stand under the shadow of the executioner's sword even as the pardon is being rushed to us from the distant capital! If its sharp blade falls, we are as dead and gone as any peasant or priest or king in the long, suffering history and prehistory of the world. Plainly, the only prudent move is to do everything possible to forestall accidental or infective death during the next decade or three, and to adopt as many healthy practices as we can manage without altogether giving up on the vivid texture of life. According to Roy Walford, for example, severe calorie restriction in your diet might eke out additional years, although this is not proved for humans and might not be the most agreeable way to plan either your dinner or your life. A brisk walk for several kilometres five days a week, on the other hand, is enjoyable, good for the heart and other muscles, lowers cholesterol and blood pressure, sheds flabby kilos, and allows you to muse on the day's activities.

That much is under the individual's control. Social and political factors play a larger role. Mortality rates vary shockingly between nations and even states or regions within nations. If we mean to ensure the best feasible health for everyone, we need to start with the basics: enough food and supplements for mothers and infants, including those in the womb, and sufficient for the rest of the community as well, clean water and air, modern sanitation, immunisation programs, decent medical and dental care. How these benefits will be funded and distributed is a topic that arouses fiery political conflict even among the generous and caring. One unexpected fact has started to become clear, however: gross inequities within even the richest nations will impact not only the poor and marginal but many of the well-off as well. It is wise insurance, at least, to make certain that people are not treated like disposable tissues—and that includes the security most of us gain by working at meaningful jobs, a declining commodity as computers extend their grasp over work.

In August 1998, *Washington Post* reporter James Lardner reviewed recent evidence suggesting that the huge income and status inequities in the USA—where one man, Bill Gates, is worth more than half the population—correlate with chronic illness and earlier death. How so? Beyond the need for a certain basic income, the key component appears to be anxiety and stress often experienced by underlings and those who feel excluded from the good life. A physiological clue is offered in baboon studies made by Stanford University neurobiologist Robert Sapolsky. Lardner notes:

> *animals lower down in the tribal hierarchy tend to resort habitually to the kind of hormonal mobilization that, in higher-status animals, is reserved for emergencies. Over time, escalated levels of cortisol and other stress-related hormones become the norm; the hippocampus gland (important to learning and memory) shrinks, and disproportionately many animals succumb to cancer, brain damage or stroke.*[21]

Drawing on similar statistics, Robin Hanson, a health policy scholar at the School of Public Health, University of California, Berkeley, has observed: 'You should pay a lot of attention to getting lots of social supports, and to reducing the chance of events that could really stress you, like crime, unemployment, divorce, etc. Don't be so focused on whether you are overweight or smoke.'[22]

If all else fails, if we are stricken before the boon is delivered, we have one last hopeful chance: the cryonics option, where following legal death (as defined by current medical criteria) you are regarded not as a disposable corpse but as a terminal patient awaiting improved treatment. The only way that can be achieved, if at all, is to perfuse the body with fluids intended to prevent more decay during this 'waiting in the ambulance' phase and then to freeze the body (or just the head) to 196 degrees C. Suspended in a well-maintained Dewar, your rigid remains will stay protected against decay for decades or even centuries, held in hope that improved medical treatments will allow revival, repair or regrowth (perhaps of the entire body) and even rejuvenation. It is a long shot, as the more cautious among cryonicists admit freely—but it beats the hell out of such traditional alternatives as incineration and rotting in the ground. Nor is it even all that expensive—a simple insurance policy covers the costs of the suspension, indefinite storage, ongoing research into recovery techniques, and a modest nest egg for your eventual return to life. Of course, it helps to live (or rather, die) near a cryonics facility, and there are not yet many of these to be found outside a few cities in the USA. I have to admit with a certain embarrassment that I am not signed up for such a program myself, although I approve of the idea as a literally last resort.

TECHNO-ABSOLUTISM?

For some, the suspicion will not be appeased that such claims, such ambitions, smack of the terrible dreams of the triumph of the will over mere matter. Time and again, those dreams fetched civilisation in the twentieth century into the hands of the morally bankrupt. That case, that renewed dread, has been expressed vividly in a recent outburst by Michael S. Malone, a contributing writer to the high-tech advocacy magazine *Upside*[23]. Malone grew up in Silicon Valley and now denounces what he sees as its own emerging triumphalist ideology, an ideology of intelligence and professional caste. 'Today we can only exclude the modestly intelligent from our companies, our neighborhoods and our private schools,' Malone declares. 'But a few years down the road when we have the right diagnostic tools—thank you, Human Genome Project—we'll be able to eliminate this burden altogether by liquidating the sub-brilliant before they're born.'

That same prospect did not trouble Lee Silver especially, you'll recall, in his projection of a growing divide between GenRich and Naturals. Malone, by contrast, sees this impulse toward transhumanist renovation as continuous with Nazi and Stalinist murders of the allegedly unfit or racially and ideologically unclean.

The risk is surely there and needs to be guarded against, but the difference, as far as I can see, is very great. It is one thing to give children the best opportunities in health and education that can be devised and funded, altogether another to stigmatise or even destroy fractions of the community. Even those who cling to the belief that abortion is murder ought to have less objection (at least on those grounds) to selecting sperm and ova, or indeed individual alleles, so as to maximise the life chances of the children conceived from those randomly generated seeds. And if techniques are developed for augmenting our powers (as effective education and nutrition already do), should we deny them to ourselves and our children? Yes indeed, says Malone. His horror is graphic:

Restructure his DNA, pop a few slivers of silicon into his cerebral cortex or just mainline his central nervous system right into the worldwide grid. Who needs neuromancy when you can hook right into the Net, become a human browser and act as your own software agent?

He ends his jeremiad in a fever close to hysteria, foreseeing 'good technofascists, genetically pure technojugend in their chip-embedded brown shirts, marching in lockstep on the Sudetenland of the computer illiterate, the unbrilliant and the imperfect. Singing songs of freedom through technology. Joyfully building the 1,000-year Digital Reich.'

Well, the moment improvement in skills, clothing, education, sensitivity or any other attribute of a human life becomes feasible, we face the risk that it will be misused to create barriers and odious discrimination. The most frightening apartheid one could imagine is a future world in which extended life is allowed only to a few—the very wealthy, the political elite and their chosen followers, nomenklatura, Mafia, military, scientists, sports heroes, movie stars, whatever. It is up to all of us to ensure that this future never happens. We will not best prevent it by denouncing technical advances and trying to blockade them, but in thinking hard, feeling deeply and wisely, debating the issues together, and acting as free men and women. If we choose to enhance the human condition, to mitigate as best we can the horrors and blunders evolution has imposed, that will be a step toward diversity in unity, one taken in the confidence that those old tyrannies were marked exactly by ferocious control and interdiction of knowledge. A deathless society without a free flow of information and opportunities to choose would indeed be hellish. We must ensure that the form of the future is not foreclosed by panic or mistrust.

A century ago, H. G. Wells looked into a future that might yet be born from the choices and aspirations of the first immortal generation. Addressing the Royal Institution in 1902, the comparatively youthful Wells (he was still only 35) spoke to his learned audience in the voice of a prophet of optimism. In the bitter irony of real history, Wells would be crushed in despair 40 years later under the torment of two global wars, the criminal playing-out in Germany, Russia, Japan and elsewhere of what critics rightly see as the totalitarian Messianism of his own utopian hopes. Yet this early vision remains vital and plausible, even inevitable, in a world where democracy is at last possible, freedom from want is genuinely in prospect, and death itself might be nearly behind us:

> It is possible to believe that all the past is but the beginning of a beginning, and all that is and has been is but the twilight of the dawn. It is possible to believe that all that the human mind has ever accomplished is but the dream before the awakening. We cannot see ... what this world will be like when the day is fully come. We are creatures of the twilight. But it is out of our race and lineage that minds will spring, that will reach back to us in our littleness to know us better than we know ourselves, and that will reach forward fearlessly to comprehend this future that defeats our eyes ... [A] day will come, one day in the unending succession of days, when beings ... who are now latent in our thoughts and hidden in our loins, shall stand upon this earth as one stands upon a footstool, and shall laugh and reach out their hands amidst the stars.[24]

A third of a century on, in his *Experiment in Autobiography*, Wells found that discourse 'vague, inexact, and rhetorical'. Quite so, yet its Edwardian optimism speaks again now to our slowly recovering sense of a deep, unplumbed future. For nearly two-thirds of a century more, civilised people have been crushed by the same despair as the ageing Wells. No Edwardian glow can compete with the chill of death camps, the flash of nuclear horror, the searing threat of a Greenhouse Effect produced (if the consensus of scientists is to be believed) by the very successes of technology.

In such a pervasive or frenetic mood of gloom, it is worth heeding the opinion of a very great scientist, Richard Feynman, uttered less than 20 years after he had helped create the atomic bomb: '... as a consequence of science one has a power to do things ... The whole industrial revolution would almost have been impossible without the development of science. The possibility today of producing quantities of food adequate for such a large population, of controlling sickness—the very fact

that there can be free men without the necessity of slavery for full production—are very likely the result of the development of scientific
means of production.'[25] Today, those means are about to remake our
very mortality.

HOPE, AT LAST

For those of us who are now alive, the signal and astonishing difference
from the immemorial past is that we really do have a chance, finally.
One might think, if an audacious analogy is required, of the freedom
fighter and political prisoner Nelson Mandela, locked away from life for
a quarter of a century—but freed at last in a triumph of democracy, and
not just freed as a wretched old man with no future but elevated into
the presidency of his new nation, sought out in many high places of the
world for his wisdom and charisma.

The twentieth century witnessed this kind of cruel oppression and
phoenix recovery again and again, most poignantly perhaps in the survivors of the Holocaust. Millions died, but some prevailed even in those
hellish conditions. Perhaps it will be that way for us, too, faced by the
impersonal and mindless cruelty of evolution's strategies of death. We
must live as vividly as we may, while life is ours—hoping that we will
triumph, in the end, even over that final enemy. And not merely pas-
sively hoping, but acting to make it so.

If we are to be the first immortal generation—the first for whom
death is not an automatic imposition of our genes—then we must
be prepared to fund the necessary research, demand that every effort be
spent on this last and most extraordinary quest, make the overthrow of
routine human mortality our epoch's responsibility and wonderful
privilege.

endnotes

notes

CHAPTER ONE

1. *Cheating Time,* W H Freeman, 1996, p 305–7.
2. Quoted, unsourced, in John J. Medina, *The Clock of Ages*, p 312. In 1998 I checked with Dr Rose on the source of this quotation and the one cited immediately below; he commented: 'I don't recall the original sources for either one. But I will stand by both of them, bearing in mind words like "could" and "substantially", which have considerable ambiguity. Indeed, Roger Gosden and I are not very far apart at all' (personal communication, 29 June 1998).
3. Pronounced 'tuh-LOM-uh-raze', by the way.
4. Pronounced 'TEL-uh-meers'.
5. *Scientific American,* January, 1997.
6. William R. Clark, *Sex and the Origins of Death* , Oxford, 1996, p 81.

CHAPTER TWO

1. *Cheating Time,* p xix.
2. Dr Freitas' *Nanomedicine* will appear in three volumes in 1999, 2000 and 2002. A useful set of frequently asked questions and answers on nanomedicine, from which this citation is drawn, can be found at: **http://www.foresight.org/Nanomedicine/NanoMedFAQ.html**
3. Cited from Blackburn's readable account at: **http://www.dist.gov.au/events/ausprize/apr98/index.html**
4. Hayflick, *How and Why We Age*, 1996, p 331.
5. By Jing Wang, Lin Ying Xie, Susan Allan, David Beach and Gregory J. Hannon, in *Genes and Development*, vol. 12, no. 12, June 15 1998.

6. See National Centre of Biotechnology site at:
   http://www.cnb.uam.es/~mblasco

7. More technically, his doctoral research topic is large-scale differential gene expression in *Saccharomyces cerevisiae* and human diploid fibroblasts.

8. S. J. Olshansky, B. A. Carnes, and D. Grahn, 'Confronting the Boundaries of Human Longevity', *American Scientist* 86(1):52-61, 1998.

9. Ibid, p 59.

10. Hagen et al., *Proceedings of the National Academy of Science*, 95:9562-6.

11. Personal communication, 20 July 1998.

12. In November 1998, the *Wall Street Journal* reported that scientists at the National Cancer Institute and elsewhere had failed to replicate Folkman's results. Immediately, others supported the original claims, while admitting to some technical problems. The approach was confirmed at Harvard by Bjorn Olsen and at Boston's Beth Israel Deaconess Medical Center. EntreMed Inc. and Bristol Myers Squibb Co. are developing endostatin and angiostatin for human trials to start in 1999. See news report at:
    http://www.cnn.com/US/9811/13/PM-CancerTreatment.ap/

13. Francis Crick, *Life Itself: Its Origins and Nature,* London, Macdonald, 1981, p 124.

14. See Aubrey D. N. J. de Grey, 'A proposed refinement of the mitochondrial free radical theory of aging,' *BioEssays*, vol. 19, no. 2 1997, pp 161-6; 'More on mitochondria and senescence', exchange between David Gershon and de Grey, *BioEssays*, vol. 19, no. 6 1997, pp 533-4; 'A Mechanism Proposed to Explain the Rise in Oxidative Stress During Aging', *Journal of Anti-Aging Medicine*, vol. 1, no. 1 1998; 'Are those 13 proteins really unimportable?', *Proceedings of the 7th International Congress of Endocytobiology*, Freiburg im Breisgau, April 5-9 1998.

15. *Nature Genetics*, June 1988, vol. 19, no. 2, pp 171–74.

16. D. B. Friedman and T. E. Johnson, 'Three mutants that extend both mean and maximum life span...', *Journal of Gerontology*, 43:B102-B109, 1988.

17. *Science*, February 24, 1994, 25:263 (5150), pp 1128–30.

18. L. Stephen Coles, *Journal of Longevity Research,* vol. 1, no. 7. His work can be explored at http://home.earthlink.net/~scoles/

19. Pronounced PLEE-uh-TROPE-ee.

20. Personal communication, July 1998.
21. Jenkins and Jenkins, p 140.
22. Ibid, p 142.
23. This remarkable article can be found at:
   http://www. businessweek.com/1998/30/covstory.htm
24. http://www.sciam.com/0197issue/0197review1.html
25. Quoted in Ricki L. Rusting, 'Why Do We Age?', *Scientific American*,
   December, 1992, p 95.
26. *Longevity Report* 58. http://www.longevb.demon.co.uk/lr.htm

CHAPTER THREE

1. *The Selfish Gene*, rev. ed., 1989, p viii.
2. Ibid, p ix.
3. *Climbing Mount Improbable*, Viking, 1996, p 254.
4. I borrow this wistful phrase from the Australian Catholic novelist
   and essayist Barry Oakley.
5. *Climbing Mount Improbable*, p 66. The examples that follow,
   including the discussion of the computer simulation done
   by Nilsson and Pelger, are well summarized in Dawkins'
   fine book.
6. *The DNA Mystique*, Freeman, 1995, p 4.
7. *Signs of Life*, Viking, 1994, p 72.
8. The key value of history is itself a message, incidentally, which de Man's
   method tends to erase, which is one of the grievous discontents many
   onlookers feel about current theory in the humanities. I discuss this
   vexed issue in my book *Theory and Its Discontents*.
9. Niles Eldredge, *Reinventing Darwin: The Great Evolutionary Debate*,
   London, Weidenfeld &Nicolson, 1995.
10. Matt Ridley, *The Red Queen: Sex and the Evolution of Human Nature*,
   Penguin, 1994, p 332.
11. John Maynard Smith, *Did Darwin Get it Right?: Essays on Games,
   Sex and Evolution*, Penguin, 1993, p 51.
12. Ibid, p 85.
13. Robert Pool, *The New Sexual Revolution*, London,
   Hodder & Stoughton, 1993, pp 140–2.
14. http://www.abcnews.com/sections/living/DailyNews/
   gender980706.html

15. *The New Sexual Revolution*, p 2.
16. Two of the chapter headings in Matt Ridley's generally impressive *The Red Queen*.
17. *New Scientist*, 3 May 1997.
18. *Darwin's Dangerous Idea*, Allen Lane, 1995, p 195.
19. Dennett's witty style is one reason I especially enjoy reading him, of all the hard-line Darwinists. It ranges fluently from friendly informality through elevated eloquence to stinging jokes – he flays his opponent Gould in a chapter entitled 'The Spandrel's Thumb', a brilliant insiderly jest that splices together Gould's book title *The Panda's Thumb*, a famous anti-adaptationist essay by Gould and geneticist Richard Lewontin, 'The Spandrels of San Marco', implying that these fellows are thumb-fingered in their argument.
20. *The DNA Mystique*, p 195.
21. *Evolution and Healing*, Weidenfelt & Nicolson, 1995, p 189.
22. Cited in Robert Wright's *The Moral Animal,* Little Brown, 1994.
23. Williams, *Plan and Purpose in Nature*, 1996, p 156.
24. See, for example, 'Gay Genes, Revisited', by John Horgan, in *Scientific American*, November 1995.
25. Nesse and Williams, 1995, p 223.
26. Ibid, p 223.
27. Ibid, p 218.
28. Howard Bloom, *The Lucifer Principle: A Scientific Expedition into the Forces of History*, Sydney, Allen & Unwin, 1995, p 2.
29. *The Moral Animal*, especially chapter 10.
30. 'Theorists See Survival Value in Menopause', *New York Times*, September 16, 1997.
31. 'Reproductive cessation in female mammals', *Nature*, 392, 807 (1998).
32. *Plan and Purpose in Nature*, p 125.
33. William R. Clark, *Sex and the Origins of Death*, p 81.
34. Hayflick, *How and Why We Age*, 1996, p 84.
35. Richard Gosden, *Cheating Time*, p 289.
36. Carl Sagan and Ann Druyan, *Shadows of Forgotten Ancestors*, Sydney, Random House, 1992, p xiii.
37. Deborah Bloom, *The Monkey Wars*, 1994, p 24.
38. William Jordan, *Divorce Among the Gulls: An Uncommon Look at Human Nature*, London, Abacus, pp 187–205.

39. Sagan and Druyan, p 414.

40. *Guns, Germs, and Steel*, Jonathan Cape, 1997, p 78.

41. George Johnson, *Fire in the Mind: Science, Faith and the Search for Order*, Viking, 1996.

42. *The Quark and the Jaguar*, Little Brown, 1994, p 199.

43. *The Collapse of Chaos*, Penguin, 1995.

44. Ibid, p 435.

45. They resemble the 'data-farms' that Rory Barnes and I predicted in 1982 in our novel *Valencies*.

CHAPTER FOUR

1. pp 35-6.

2. This protracted debate with Chalmers, and others such as Dennett is collected in John R. Searle, *The Mystery of Consciousness*, London, Granta, 1997.

3. A popular version of Calvin's case is *How Brains Think*, Weidenfeld & Nocolson Science Masters, 1996. A more technical account is *The Cerebral Code*, MIT, 1996.

4. David J. Chalmers, *The Conscious Mind: In Search of a Fundamental Theory*, Oxford University Press, 1996, p 305.

5. Ian Stewart and Jack Cohen, *Figments of Reality*, 1997, p 211.

6. Cited by Semir Zeki, in *A Vision of the Brain*, Oxford, Blackwell, 1993.

7. *Consciousness Explained*, Allen Lane, 1992, p 37.

8. This 'colour phi phenomenon' is discussed in Dennett, *Consciousness Explained*, pp 120–6.

9. Ibid, p 418.

10. Ibid, pp 428–9.

11. Lewis Thomas, *The Fragile Species*, New York, Scribner's, 1992, p 192.

12. Nicholas Humphrey, *A History of the Mind*, London, Chatto & Windus, 1992, p 207.

13. Searle quoted as epigraph to Francis Crick, *The Astonishing Hypothesis*, 1994, p vii.

14. 'Faster than a speeding brain', *New Scientist*, 20 April 1996.

15. His World Wide Web site is: **http://www.hip.atr.co.jp/~degaris**

16. Updated information is on de Garis' Web site, the source of this quote.

17. Interview with M. Mitchell Waldrop, in *Complexity: The Emerging Science at the edge of Order and Chaos*, p 209.

18. Ibid, p 210.

19. Alfred Bester, *Tiger! Tiger!*, Penguin, 1967, p 227.

20. Richard E. Cytowic, *The Man Who Tasted Shapes*, London, Abacus, 1994, p 4.

21. Michael S. Gazzaniga, *Nature's Mind: The Biological Roots of Thinking, Emotions, Sexuality, Language and Intelligence*, Penguin, 1994.

22. William H. Calvin and Derek Bickerton, *Lingua ex Machina: Reconciling Darwin and Chomsky with the human brain*, chapter 16 (MIT Press, forthcoming 1999).

23. Gerald Edelman's difficult treatises on topobiology are usefully summarized in his *Brilliant Air, Brilliant Fire*, Penguin, 1992.

24. Gazzaniga, p 124.

25. Steven Pinker, *The Language Instinct*, 1994, p 210. The diabolical sentence was devised by Pinker's student Annie Senghas (p 209).

26. Alan Cromer, *Uncommon Sense: The Heretical Nature of Science*, Oxford University Press, 1993, p 188.

27. *The Language Instinct*, p 351.

28. Sue Savage-Rumbaugh and Roger Lewin, *Kanzi: The Ape at the Brink of the Human Mind*, London, Doubleday, 1994.

29. Antonio Damasio, *Descartes' Error*, Picador, 1994, p xiv.

30. John McCrone, *The Myth of Irrationality*, Macmillan, 1993, pp 165–87.

31. Cited with permission from a posting to the Extropian list on the Internet.

32. Roger Penrose, *The Emperor's New Mind*, Oxford University Press, 1989, and *Shadows of the Mind*, Oxford University Press, 1994.

CHAPTER FIVE

1. p 210.

2. p 263.

3. Hans Eysenck, 'Little Hans or Little Albert,' in *Fact and Fiction in Psychology*, Pelican, 1965, pp 95–131.

4. Discussed in detail in Walter Moore, *Schrödinger: Life and Thought*, Cambridge University Press, 1989.

5. Black holes, incidentally, were mentioned in 1935 in E. R. Eddison's baroque fantasy *Mistress of Mistresses*, the first of his sequels to *The Worm Ouroboros*. This struck me as pretty neat for a British civil servant

thirtysomething years before Wheeler's version of the blackhole:
> But now, just as (they tell us) a star of earthly density but of the size of
> Betelgeuze would of necessity draw to it not matter and star-dust only
> but the very rays of imponderable light, and suck in and swallow at
> last the very boundaries of space into itself, so all things condensed in
> her as to a point.

True, Karl Schwarzschild had discovered black hole singularities
as a prediction in Einstein's general relativity in 1915, two decades
earlier, but it had been my impression that the news hadn't actually hit
the headlines. I mentioned this to Paul Davies, who informed me that
singularities were the stuff of learned gossip at just that time … Bad luck.

6. *The Quark and the Jaguar,* Hawking is cited p 153.

7. David Hodgson, *The Mind Matters: Consciousness and Choice in a Quantum World*, Oxford University Press, 1991.

8. Ibid, p 461.

9. *The Quark and the Jaguar*, p 153.

10. Ibid, p 165.

11. Michio Kaku, *Hyperspace*, Oxford University Press, 1994.

12. Kip S. Thorne, *Black Holes and Time Warps: Einstein's Outrageous Legacy*, London, Picador, 1994.

13. Dean I. Radin, *The Conscious Universe: The Scientific Truth of Psychic Phenomena*, New York, HaperEdge, 1997.

14. Steven Weinberg, *Dreams of a Final Theory*, London, Hutchinson Radius, 1993, p 138.

15. Ibid, p 200.

16. Ibid, p 42.

17. Paul Davies, ed, *The New Physics*, Cambridge University Press, 1992, p 1.

18. Ibid, Chris Isham, 'Quantum gravity', p 93.

19. David Layzer, *Cosmogenesis: the Growth of Order in the Universe*, Oxford University Press, 1990.

20. Fred Hoyle, *The Intelligent Universe*, New York, Holt, Rhinehart and Winston, 1980.

21. John Archibald Wheeler, *At Home in the Universe*, American Institute of Physics, 1994.

22. Fred Alan Wolf, *Mind and the New Physics*, London, Heinemann, 1984, p 153.

23. Heinz Pagels, *The Cosmic Code: Quantum Physics as the Language of Nature*, Pelican, 1984, p 160.

24. Ibid, p 165.

25. Abraham Pais, *Niels Bohr's Times, in Physics, Philosophy, and Polity*, Oxford, Clarendon Press, 1991, p v.

26 Pais was making fun of Gary Zugav's New Age best-seller, *The Dancing Wu Li Masters: An Overview of the New Physics*, Hutchinson, 1979.

27. *Dreams of a Final Theory*, p 41.

28. p xxi, p 52, p. 219.

29. While that last claim might sound like a preposterous slur, actually it simply notes an ecological tragedy of immense proportions, one that no early human culture could possibly have expected or averted. During the last 35,000 years, according to mammal specialist Tim Flannery in *The Future Eaters* (1994), southern Africa has lost five per cent of its large mammals, Europe less than a third, North America 73 per cent, and Australia a staggering 94 per cent. Was this due to climatic shifts and depredations? Unlikely. There have been seventeen ice ages in the last two million years, and only the most recent overlapped the great Australian extinctions. What's more, these dates of the great dyings correlate all too eerily with the move of canny humans out of Africa and into new lands. The big beasts in our African birthplace, and even in much of Europe, had time to grow wary of humans, time for evolution to select for life-saving caution. Those innocent creatures discovered in America 11,000 years ago, and very belatedly in Australia (later still in New Zealand), were shockingly easy prey. They died and died. And as they died, the coevolved ecologies of plant and animal warped and broke. In Australia, Flannery remarks, more than 50 medium and large herbivorous marsupials thrived 100,000 years ago, along with giant birds and turtles. They cropped the undergrowth, swiftly returned nutrients to the already poor soil, and stabilised the ecology. When humans arrived and slew and ate the megafauna, the undergrowth went wild. Fire exploded. Soils collapsed and blew away. A few plant species adapted to fire in the dry heathlands spread out and colonised the suddenly denuded landscapes. The newly arrived humans we now call Aborigines were caught up in the long disaster, and did what they could to manage its impact by deliberately burning patches

of land, a technological development called 'firestick farming'.
Their arrival had wrecked and remade the country, but their inventive
response 'prevented the establishment of the vast fires that stripped soils
and nutrients most dramatically,' notes Flannery. 'They also prevented the
loss of small remnant patches of rainforest ...' The much later arrival
of Europeans, with their ignorance of this specialised ecology, their
superior mechanical knowledge, and their hunger for new farmland,
'was to prove the undoing of 35,000 years of conservation efforts.'

30. Michael Allaby, *Facing the Future: The Case for Science*,
London, Bloomsbury, 1995, p 60.

31. Ibid, p 127.

CHAPTER SIX

1. Vernor Vinge, *A Fire Upon the Deep*, London, Millennium, p 1.

2. George Gamow, *The Creation of the Universe*, New York, Mentor, 1957.

3. Fred Hoyle, *Frontiers of Astronomy*, London, Heinemann, 1955.

4. *The Creation of the Universe*, p 48.

5. Nothing daunted, he also applied his theory to evolution – 'beta genes',
he suggests, both provoke and edit mutations, spurring development –
and to the nature of language.

6. Eric Lerner, *The Big Bang Never Happened*, New York, Vintage,
rev. ed., 1992.

7. George Smoot, *Wrinkles in Time: The Imprint of Creation*,
London, Little Brown, 1993, p 289.

8. *Superforce*, p 10.

9. John Barrow, *Impossibility*, Oxford, 1998, p 66.

10. Ibid, p 67.

11. Steven Weinberg, *The First Three Minutes*, London, Deutsch, 1977.
Recalling this celebrated statement in his 1993 book *Dreams
of a Final Theory*, Weinberg ruefully remarks that 'I did not mean
that science teaches us that the universe is pointless, but rather that
the universe itself suggests no point. I hastened to add that there were
ways that we ourselves could invent a point for our lives...' (p 204).

12. Personal communication, 5 October 1998.

13. Ian Stewart, *Nature's Numbers: Discovering Order and Pattern
in the Universe*, Weidenfeld & Nicolson, 1995, p 39.

14. Ibid, p 147.

15. This is a recurrent theme in the many excellent popular books by Paul Davies, and is explored at some length in Chapter 4, 'Symmetry and beauty', in *Superforce*, pp 50–69.

16. John Barrow, *Pi in the Sky: Counting, Thinking, and Being*, Clarendon Press, Oxford, 1992, p 296.

17. Margaret Wertheim, *Pythagoras' Trousers*, Times Books, Random House, 1995, p 233.

18. Ibid, p 15.

19. Ibid, p 251.

20. Ibid, p 195.

21. Arthur Koestler, *The Act of Creation*, London, Hutchinson, 1964, and other books.

22. Paul Davies, *The Fifth Miracle*, p 227.

23. Ibid, p 215.

24. Frank Tipler has made this argument in several places, notably in Barrow and Tipler, *The Anthropic Cosmological Principle*.

25. See Lee Smolin's *The Life of the Cosmos*, and Stephen Hawking's *Black Holes and Baby Universes*, London, Bantam Press, 1993.

26. John Gribbin, *In the Beginning: The Birth of the Living Universe*, London, Viking, 1993, p 208.

27. Australian commentator Mitch Porter remarked, when I raised this objection, 'Yes, the parameter variation between generations couldn't be too great. I can think of one reason why this might be so: if Grand Unified Theory symmetry-breaking is any model for Smolin's parameter change, then parameter change consists of moving from one local minimum to another on a potential surface in parameter space. In general, the greater the change, the more energy required' (personal communication).

28. David Deutsch, *The Fabric of Reality*, Allen Lane, 1997.

29. Ibid, p 45.

30. http://www.sns.ias.edu/~max/toe.html

31. 'Is "the theory of everything" merely the ultimate ensemble theory?', preprint, 3 April 1997, p 27. I am most grateful to Dr Tegmark for making this paper available to me, as well as other papers: 'On the dimensionality of spacetime' (accepted for publication in *Classic and Quantum Gravity*); and 'Does the universe in fact contain almost no information', *Foundations of Physics Letters*, vol. 9, no. 1, 1996.

CHAPTER SEVEN

1. Fossel, *Reversing Human Aging*, W Morrow, 1996, p 4.
2. Hofstadter, p 481.
3. Michael Binyon, *The Times* (in the *Weekend Australian*, 11/12 July 1998, p 15).
4. Dr George D. Moffett, *Critical Masses*.
5. Cited with permission from a list on the Internet.
6. http:///www.inform.umd.edu/EdRes/Colleges/ BMGT/.Faculty/Jsimon/Ultimate-Resource/
7. *Holy Fire*, p 47.
8. Ibid, p 48
9. Ibid, p 52.
10. Ibid.
11. See the crisp discussion of this political likelihood, Ibid, pp 49–51.
12. Cited with permission from an Internet newsgroup posting.
13. In George Turner, *A Pursuit of Miracles*, Adelaide, Aphelion, 1990.
14. 1998 CD-ROM edition, in the article 'Kinds of Insurance: Life and Health Insurance'.
15. David Brin, *The Transparent Society: Will Technology Force Us to Choose Between Privacy and Freedom?*, Perseus Press, 1998.
16. This citation is drawn from: http://www.foresight.org/Nanomedicine/NanoMedFAQ.html
17. Personal communication, June 1998.
18. Cited from Professor Imhof's Internet site at: http://userpage.fuberlin.de/~aeimhof/
19. *Reversing Human Aging*, p 222. Dr Fossel's book, although already somewhat dated by laboratory advances, is still perhaps the best general treatment of the likely impact of telomere and similar life-extension treatments. It is appealingly modest, for all its Promethean claims, and warmly humane. To my surprise, when I sought Elizabeth Blackburn's expert opinion of Fossel's arguments, she said she had not read the book.
20. Vol. 62, no. 11. Skulachev is the journal's editor-in-chief.
21. James Lardner, 'Deadly disparities: Americans' widening gap in incomes may be narrowing our lifespans', *Washington Post*, 16 August 1998, p C01 (Outlook Section).
22. Hanson made this comment, based on recent multifactorial health survey results, on the Extropian email list, 10 August 1998.

23. The following citations are from the Internet e-magazine *Upside*, June 29, 1998, at **http://www.upside.com**

24. H. G. Wells, 'The Discovery of the Future,' delivered at the Royal Institution on January 24, 1902. Printed in *Nature*, 65, 326, February 6, 1902.

25. Richard Feynman, *The Meaning of It All: The 1963 John Danz lectures*, Allen Lane, 1998, p 4.

recommended reading

LIFE

Per Bak, *How Nature Works: The Science of Organized Criticality*,
 Oxford University Press, 1997

William R. Clark, *At War Within: The Double-Edged Sword of Immunity*,
 Oxford University Press, 1995, and *Sex and the Origins of Death*,
 Oxford University Press, 1996

Peter Coveney and Roger Highfield, *Frontiers of Complexity:
 The Search for Order in a Chaotic World*, Faber, 1995

Paul Davies, *The Fifth Miracle: The Search for the Origin of Life*, Allen Lane, 1998

Richard Dawkins, *Climbing Mount Improbable*, Viking, 1996

Daniel C. Dennett, *Darwin's Dangerous Idea: Evolution and the
 Meanings of Life*, Allen Lane, 1995

Jared Diamond, *Guns, Germs and Steel: A Short History of Everybody
 for the Last 13,000 Years*, Jonathan Cape, 1997

Stephen Jay Gould, *Wonderful Life: The Burgess Shale and the Nature
 of History*, Penguin, 1991, and *Life's Grandeur*, Jonathan Cape, 1997

George Johnson, *Fire in the Mind: Science, Faith and the Search for Order*,
 Viking, 1995

Stuart A. Kauffman, *The Origins of Order: Self-Organization and Selection
 in Evolution*, Oxford University Press, 1993

Roger Lewin, *Complexity: Life at the Edge of Chaos*, Dent, 1993

Dorothy Nelkin and M. Susan Lindee, *The DNA Mystique: The Gene
 as a Cultural Icon*, Freeman, 1995

Randolph M. Nesse and George C. Williams, *Evolution and Healing:
 The New Science of Darwinian Medicine*, Weidenfeld & Nicolson, 1995

Robert Pollack, *Signs of Life: The Language and Meanings of DNA*, Viking, 1994

Robert Pool, *The New Sexual Revolution,* Hodder Stoughton, 1993

Lee M. Silver, *Remaking Eden: Cloning and Beyond in a Brave New World*, Weidenfeld & Nicolson, 1998

M. Mitchell Waldrop, *Complexity: The Emerging Science at the Edge of Order and Complexity,* Viking, 1992

George C. Williams, *Plan & Purpose in Nature,* Weidenfeld & Nicolson, 1996

Edward O. Wilson, *Naturalist,* Allen Lane, 1994, and *Consilience: The Unity of Knowledge*, Little Brown, 1998

Robert Wright, *The Moral Animal: The New Science of Evolutionary Psychology*, Little, Brown, 1994

MIND

Margaret Boden, *The Creative Mind*, rev. ed, Cardinal, 1992

William H. Calvin, *How Brains Think: Evolving Intelligence, Then and Now*, Weidenfeld & Nicolson, 1996, and *The Cerebral Code: Thinking a Thought in the Mosaics of the Mind*, MIT Press, 1996

William H. Calvin and Derek Bickerton, *Lingua ex Machina: Reconciling Darwin and Chomsky with the human brain*, MIT Press, forthcoming, 1999

David J. Chalmers, *The Conscious Mind: In Search of a Fundamental Theory*, Oxford University Press, 1996

Antonio R. Damasio, *Descartes' Error: Emotion, Reason and the Human Brain*, Picador, 1994

Daniel C. Dennett, *Consciousness Explained*, Allen Lane, 1992

Gerald Edelman, *Brilliant Air, Brilliant Fire: On the Matter of Mind*, Allen Lane, 1992

Howard Gardner, *The Mind's New Science: A History of the Cognitive Revolution*, Basic Books, 1985

Douglas R. Hofstadter, *Gödel, Escher, Bach: An Eternal Golden Braid*, Penguin, 1980, *Metamagical Themas: Questing for the Essence of Mind and Pattern,* Viking, 1985, and *Le Ton beau de Marot: In Praise of the Music of Language*, Basic Books, 1998

Douglas R. Hofstadter, and Daniel C. Dennett, eds, *The Mind's I: Fantasies and Reflections on Self and Soul*, Penguin, 1982

Jerome Kagan, *Galen's Prophecy: Temperament in Human Nature*,

Basic Books, 1994

John McCrone, *The Myth of Irrationality: The Science of the Mind from Plato to Star Trek*, Macmillan, 1993

Steven Mithen, *The Prehistory of the Mind: A Search for the Origins of Art, Religion and Science*, Thames and Hudson, 1996

Steven Pinker, *The Language Instinct*, Allen Lane, 1994, and *How the Mind Works*, Allen Lane, 1997

Ian Stewart and Jack Cohen, *Figments of Reality: The Evolution of the Curious Mind*, Cambridge University Press, 1997

Semir Zeki, *A Vision of the Brain*, Blackwell, 1993

QUANTUM

John Barrow, *Theories of Everything: The Quest for Ultimate Explanation*, Oxford University Press, 1991, and *Impossibility: The Limits of Science and the Science of Limits,* Oxford University Press, 1998

Jack Cohen and Ian Stewart, *The Collapse of Chaos: Discovering Simplicity in a Complex World*, Penguin, 1995

David Deutsch, *The Fabric of Reality*, Allen Lane, 1997

Murray Gell-Mann, *The Quark and the Jaguar: Adventures in the Simple and the Complex*, Little, Brown, 1994

John Gribbin, *Schrödinger's Kittens and the Search for Reality*, Weidenfeld & Nicolson, 1995

Gerard Milburn, *Quantum Technology*, Allen & Unwin, 1996, and *The Feynman Processor: An Introduction to Quantum Computation*, Allen & Unwin, 1998

Heinz R. Pagels, *The Cosmic Code: Quantum Physics as the Language of Nature*, Pelican, 1984

Lewis Wolpert, *The Unnatural Nature of Science*, Faber, 1992

COSMOS

John Barrow and Frank Tipler, *The Anthropic Cosmological Principle*, Oxford University Press, 1986

Paul Davies, *The Cosmic Blueprint: Order and Complexity at the Edge of Chaos*, Heinemann, 1987, *The Mind of God: The Scientific Basis for a Rational World*, Simon and Schuster, 1992, and *About Time: Einstein's Unfinished Revolution*, Viking, 1995

Timothy Ferris, *Coming of Age in the Milky Way*, Vintage, 1988, and *The Whole Shebang: A State-of-the-Universe(s) Report*,

Weidenfeld & Nicolson, 1997

John Gribbin, *Companion To The Cosmos*, Weidenfeld & Nicolson, 1996

Dennis Overbye, *Lonely Hearts of the Cosmos*, HarperCollins, 1991

Lee Smolin, *The Life of the Cosmos*, Weidenfeld & Nicolson, 1997

Kip S. Thorne, *Black Holes and Time Warps: Einstein's Outrageous Legacy*, Picador, 1994

Frank Tipler, *The Physics of Immortality: Modern Cosmology, God and the Resurrection of the Dead*, Macmillan, 1995

Margaret Wertheim, *Pythagoras' Trousers: God, Physics, and the Gender Wars*, Times Books, Random House, 1995

John Archibald Wheeler, *At Home in the Universe*, American Institute of Physics, 1994

Defeating Ageing and Death

Ben Bova, *Immortality: How Science is Extending Your Life– and Changing the World*, Avon, New York, 1998

Damien Broderick, *The Spike: Accelerating into the Unimaginable Future*, Reed Books/New Holland, 1997

Arthur C. Clarke, *Profiles of the Future*, Pan Books (revised ed), 1983

K. Eric Drexler, *Engines of Creation: The Coming Era of Nanotechnology*, Doubleday Anchor, 1986, and *Nanosystems: Molecular Machinery, Manufacturing, and Computation*, John Wiley, 1992

Robert C. W. Ettinger, *The Prospect of Immortality*, New York, Doubleday, 1964 (available on the Web at: **http://www.cryonics.org/book1.html**), and *Man into Superman*, New York, St. Martin's Press, 1972 (available on the Web at: **http://www.cryonics.org/book2.html**)

Michael Fossel, *Reversing Human Aging*, William Morrow, 1996

Roger Gosden, *Cheating Time: Science, Sex, and Ageing*, W. H. Freeman, 1996

Hugh Hixon, 'Misadventure as a Cause of Death in an Immortal Population' *Cryonics*, May 1998

Leonard Hayflick, *How and Why We Age*, Ballantine Books, 1996

Susan Jenkins and Robert Jenkins, *Life Signs: The Biology of Star Trek*, HarperCollins, 1998

Michio Kaku, *Visions: How Science Will Revolutionize the 21st Century and Beyond*, Oxford University Press, 1998

John J. Medina, *The Clock of Ages: Why We Age, How We Age,*

Winding Back the Clock, Cambridge University Press, 1996

Hans Moravec, *Mind Children: The Future of Robot and Human Intelligence*, Harvard University Press, 1988, and In *Robot: Mere Machine to Transcendent Mind*, Oxford University Press, 1998

Ed Regis, *Great Mambo Chicken and the Transhuman Condition*, Viking, 1991, and *Nano! Remaking the World Atom by Atom*, Bantam, 1995

David W. E. Smith, *Human Longevity*, Oxford University Press, 1993

Bruce Sterling, *Holy Fire*, Millennium, 1996

acknowledgments

Thanks to the people who provided the information I have tried to distil in this book. I am sure to have garbled or distorted at least some of what I have gathered from these learned men and women, but I hope they will forgive me for that (and let me know, so I don't do it again). I am grateful to Douglas Hofstadter for allowing me, despite his misgivings, to quote his grieving words as epigraph to chapter 7. Much of the research material I discuss is drawn from the books listed above, and from many others as well, and from journal and newspaper articles and, especially, from the wonderful hoard of knowledge now available on the Internet. I am especially grateful to research workers in fields that one day will give us extended and perhaps indefinite life. Many of them have answered my questions, or sent me to the latest sources of information.

Thanks to Dr Woodring Wright, Dr Elizabeth Blackburn, Dr Michael Rose, and others working on telomeres and related puzzles. A non-scientist, Thomas Mahoney, who founded Lifeline Laboratories, Inc. in 1996, provided a strong argument and many references in favour of the telomeric theory. I am greatly indebted to Aubrey de Grey for sending me his papers on the mitochondrial free radical theory of ageing, and especially for taking the trouble to read the relevant chapters of this book and provide detailed commentary; while usually I have accepted his corrections, and in one place borrowed a paragraph wholesale, he is of course not responsible for any misunderstandings that persist here.

Many other people have taken a few minutes to answer my urgent and sometimes ignorant emails. I am most indebted, as well, to Russell Blackford, who read the whole manuscript and made so many detailed and insightful comments that I started again from the top. Thanks also to those who read fragments of the work in progress, and sent me useful corrections, quotations, URLs, jokes and good ideas I'd never have thought of by myself: people like Hugh Hixon, Chris Lawson, Kathryn Aegis, Robin Hanson, and others whose names you will find scattered through my pages. Not all agree with my opinions; some will be vehemently opposed. I appreciate their contributions.

Race Mathews and Victor Perton, MP, provided useful bipartisan support and agreeable luncheons at Parliament House.

My editor and my publisher at New Holland have been kindly, efficient, and supportive. Without the initial interest of my Reed/New Holland commissioning editor and publisher Clare Coney, who earlier published *The Spike*, I might never have started this complex book.

As always, I am grateful to the University of Melbourne, where I am a Fellow of the Department of English and Cultural Studies, for their generous support in providing the email and Net links, photocopying and library access that an independent scholar needs so desperately.

And I am especially indebted to the people of the state of Victoria, for a grant from their Arts Development Program. May they all live forever!

Melbourne, January 1999

index

index